QUANTITATIVE TECHNIQUES

QUANTITATIVE TECHNIQUES

Dr. Sudhir Gupta

CENTRUM PRESS
NEW DELHI-110002 (INDIA)

CENTRUM PRESS
H.O.: 4360/4, Ansari Road, Daryaganj,
New Delhi-110002 (India)
Tel: 23278000, 23261597, 23255577, 23286875
B.O.: No. 1015, Ist Main Road, BSK IIIrd Stage,
IIIrd Phase, IIIrd Block, Bangalore-560085 (INDIA)
Tel: 080-41723429
Email: centrumpress@gmail.com
Visit us at: www.centrumpress.com

Quantitative Techniques

First Edition, 2011

ISBN 978-93-80921-14-3

PRINTED IN INDIA

Printed at Balaji Offset, Delhi.

Contents

Preface

Management education in India gained momentum only in the sixties. In the early stages, faculty, concepts, materials and pedagogy were borrowed from the Western collaborators. Soon it was realised that while the basic concepts, tool and techniques of management might be common, research had to be done in many areas of contextually relevant teaching materials had to be developed. The need for contextual relevance of the application of quantitative techniques to developing countries has been increasingly felt because management principles are not only being applied to business and industry, but also to newer areas like rural development, non-profit making organisations and non-enterprise management.

Our objective with quantitative techniques is to provide a textbook for Punjab Technical University that comprehensively explores. The discussion made in the book will help the readers to understand the different and unique aspects of "*Quantitative Techniques*".

Author

Preface

1

Introduction

CONCEPT OF QUANTITATIVE TECHNIQUES

Empirical analysis lies at the heart of modern competition policy. The outcomes of merger cases, dominance claims and the assessment of potentially anti-competitive agreements all depend on the accurate analysis and interpretation of empirical evidence.

This Lexecon Ltd publication provides an overview of the wide range of quantitative techniques that are available for use in competition inquiries.

THE ROLE OF QUANTITATIVE TECHNIQUES

It is sometimes suggested that quantitative techniques require high quality data which are usually unavailable. A related observation which is often heard is that the better the data quality, the more obscure and complex is the analysis. In reality, both concerns are overstated.

The case for seeking to carry out empirical analysis remains compelling for the following reasons. First, as this publication will emphasize, many quantitative techniques can be successfully executed using fairly rudimentary data. Many of the available techniques can produce highly effective results with basic information.

The intuitive and accessible nature of these techniques means that they can be understood by a wide audience. Second, it is undeniable that some of the more sophisticated techniques raise complicated technical issues and require considerable expertise on the part of the practitioner. However, precisely because of their sophistication, these techniques can yield exact answers to the questions that are pivotal to competition cases.

Competition authorities are increasingly well versed in these areas. The European Commission, for example, is now familiar with both merger simulation and econometric analysis. It is clear that the role of the more complex approaches outlined in this publication will continue to expand.

WHICH TECHNIQUE

The following quantitative techniques are covered in this publication.

Price Correlation Analysis

This examines the extent to which the prices of two products move together over time. If the price of one good constrains the price of the other, the two price series follow a similar pattern.

Stationarity Analysis

This is a more sophisticated variant of price correlation analysis. Stationarity analysis does not suffer from some of the complications associated with price correlation analysis.

Price Elasticity Analysis

This uses econometrics to assess the responsiveness of the demand for a product or group of products to a change in price. The more elastic the demand, the less likely is it that a price rise would be profitable.

Critical Loss Analysis

This makes the SSNIP market definition test operational by estimating how much the hypothetical monopolist's sales would have to fall in order to make the hypothesised price increase unprofitable.

Switching Analysis

This uses econometric analysis, internal company documentary evidence or survey results to estimate the extent to which two products are particularly 'close' competitors.

Merger Simulation

This combines estimates of demand elasticities with an assumption about the nature of competition to simulate the competitive impact of a proposed merger.

Price/Concentration Analysis

This explores the degree to which higher levels of concentration coincide with higher prices and margins in order to give an insight into the likely competitive impact of an increase in market concentration.

Shock Analysis

This involves the analysis of discrete events affecting competition in the market to obtain an insight into the competitive interaction between products or regions.

Bidding Studies

In some markets, competition takes the form of bidding for contracts. The analysis of bidoutcomes can identify the extent to which merging firms compete with each other.

Damages Assessment

It is possible to quantify the impact of anti-competitive behaviour using statistical techniques; the assessment of damages associated with price fixing to show what can be achieved. The choice of technique will depend on both the specific circumstances of the case and the availability of data. Also, it is possible that more than one technique could be usefully applied to any particular case.

ROLE OF MATHEMATICS IN BUSINESS DECISION

The role of mathematics in Business decisions has very

important in the process of mangerial decision models and alorithms. To turn to the specific aspects of te quantitative decision making process, it is possible to recognize three distinct phases in every decision situation. First is carefully defined the problem, second is a conceptual model to be generated and third is the selection of the appropriate quantitative model thay may lead to a solution. lastly a specific alogorithm is selected. An alorithms are the ordely delineated sequnces of mathematical operations that lead to a solution. The alorithms generate the decision which is subsequently implemented managerial action programme.:

Defined problem → Conceptual model → Quntitative Model → Algorithms → decision – Action programmes

STATISTICS IN BUSINESS DECISIONS

Business statistics is the science of good decision making in the face of uncertainty and is used in many disciplines such as financial analysis, econometrics, auditing, production and operations including services improvement, and marketing research. This makes the topic of time series especially important for business statistics. It is also a branch of applied statistics working mostly on data collected as a by-product of doing business or by government agencies. It provides knowledge and skills to interpret and use statistical techniques in a variety of business applications. A typical business statistics course is intended for business majors, and covers statistical study, descriptive statistics, probability, and the binomial and normal distributions, test of hypotheses and confidence intervals, linear regression, and correlation.

THEORY OF SETS

Set theory is the branch of mathematics that studies sets, which are collections of objects. Although any type of object can be collected into a set, set theory is applied most often to objects that are relevant to mathematics.

The modern study of set theory was initiated by Georg Cantor and Richard Dedekind in the 1870s. After the discovery of paradoxes in naive set theory, numerous axiom systems were proposed in the early twentieth century, of which the Zermelo-Fraenkel axioms, with the axiom of choice, are the best-known.

The language of set theory is used in the definitions of nearly all mathematical objects, such as functions, and concepts of set theory are integrated throughout the mathematics curriculum. Elementary facts about sets and set membership can be introduced in primary school, along with Venn and Euler diagrams, to study collections of commonplace physical objects. Elementary operations such as set union and intersection can be studied in this context. More advanced concepts such as cardinality are a standard part of the undergraduate mathematics curriculum.

Set theory is commonly employed as a foundational system for mathematics, particularly in the form of Zermelo-Fraenkel set theory with the axiom of choice. Beyond its foundational role, set theory is a branch of mathematics in its own right, with an active research community. Contemporary research into set theory includes a diverse collection of topics, ranging from the structure of the real number line to the study of the consistency of large cardinals.

BACKDROP

Mathematical topics typically emerge and evolve through interactions among many researchers. Set theory, however, was founded by Georg Cantor:"On a Characteristic Property of All Real Algebraic Numbers".

Since the 5th century BC, beginning with Zeno in the West and early Indian mathematicians in the East, there had been mathematicians struggling with the concept of infinity. The Indian Jains in the 4th century BC proposed the concept of there being different types of infinities: infinite in length, infinite in area, infinite in volume, and infinite perpetually. In the 10th century AD, Udayana, founder of the Navya-Nyaya school of Indian logic, developed theories on"restrictive conditions for universals" and"infinite regress" that anticipated aspects of modern set theory. Especially notable is the work of Bernard Bolzano in the first half of the 19th century. The modern understanding of infinity began in 1867-71, with Cantor's work on number theory. An 1872 meeting between Cantor and Richard Dedekind influenced Cantor's thinking and culminated in Cantor's 1874.

Cantor's work initially polarized the mathematicians of his day. While Karl Weierstrass and Dedekind supported Cantor, Leopold Kronecker, now seen as a founder of mathematical constructivism, did not. Cantorian set theory eventually became widespread, due to the utility of Cantorian concepts, such as one-to-one correspondence among sets, his proof that there are more real numbers than integers, and the"infinity of infinities" the power set operation gives rise to.

The next wave of excitement in set theory came around 1900, when it was discovered that Cantorian set theory gave rise to several contradictions, called antinomies or paradoxes. Bertrand Russell and Ernst Zermelo independently found the simplest and best known paradox, now called Russell's paradox and involving"the set of all sets that are not members of themselves." This leads to a contradiction, since it must be a member of itself and not a member of itself. In 1899 Cantor had himself posed the question:"what is the cardinal number of the set of all sets?" and obtained a related paradox.

The momentum of set theory was such that debate on the paradoxes did not lead to its abandonment. The work of Zermelo in 1908 and Abraham Fraenkel in 1922 resulted in the canonical axiomatic set theory ZFC, which is thought to be free of paradoxes. The work of analysts such as Henri Lebesgue demonstrated the great mathematical utility of set theory. Axiomatic set theory has become woven into the very fabric of mathematics as we know it today.

BASIC CONCEPTS

Set theory begins with a fundamental binary relation between an object *o* and a set *A*. If *o* is a member of *A*, we write *o* " *A*. Since sets are objects, the membership relation can relate sets as well.

A derived binary relation between two sets is the subset relation, also called set inclusion. If all the members of set *A* are also members of set *B*, then *A* is a subset of *B*, denoted *A* " *B*. For example, {1,2} is a subset of {1,2,3}, but {1,4} is not. From this definition, it is clear that a set is a subset of itself; in cases where one wishes to avoid this, the term proper subset is defined to exclude this possibility.

Just as arithmetic features binary operations on numbers, set theory features binary operations on sets.

The:

- Union of the sets *A* and *B*, denoted *A* *" *B*, is the set of all objects that are a member of *A*, or *B*, or both. The union of {1, 2, 3} and {2, 3, 4} is the set {1, 2, 3, 4}.
- Intersection of the sets *A* and *B*, denoted *A*)" *B*, is the set of all objects that are members of both *A* and *B*. The intersection of {1, 2, 3} and {2, 3, 4} is the set {2, 3}.
- Complement of set *A* relative to set *U*, denoted A^c, is the set of all members of *U* that are not members of *A*. This terminology is most commonly employed when *U* is a universal set, as in the study of Venn diagrams. This operation is also called the set difference of *U* and *A*, denoted $U \setminus A$. The complement of {1,2,3} relative to {2,3,4} is {4}, while, conversely, the complement of {2,3,4} relative to {1,2,3} is {1}.
- Symmetric difference of sets *A* and *B* is the set of all objects that are a member of exactly one of *A* and *B*. For instance, for the sets {1,2,3} and {2,3,4}, the symmetric difference set is {1,4}.
- Cartesian product of *A* and *B*, denoted $A \times B$, is the set whose members are all possible ordered pairs (a,b) where *a* is a member of *A* and *b* is a member of *B*.
- Power set of a set *A* is the set whose members are all possible subsets of *A*. For example, the powerset of {1, 2} is {{}, {1}, {2}, {1,2}}.

SET (MATHEMATICS)

A set is a collection of distinct objects, considered as an object in its own right. Sets are one of the most fundamental concepts in mathematics. Although it was invented at the end of the 19th century, set theory is now a ubiquitous part of mathematics, and can be used as a foundation from which nearly all of mathematics can be derived. In mathematics education, elementary topics such as Venn diagrams are taught at a young age, while more advanced concepts are taught as part of a university degree.

ALGEBRA OF SETS

The algebra of sets develops and describes the basic properties and laws of sets, the set-theoretic operations of union, intersection, and complementation and the relations of set equality and set inclusion. It also provides systematic procedures for evaluating expressions, and performing calculations, involving these operations and relations.

The algebra of sets is the development of the fundamental properties of set operations and set relations. These properties provide insight into the fundamental nature of sets. They also have practical considerations.

Just like expressions and calculations in ordinary arithmetic, expressions and calculations involving sets can be quite complex. It is helpful to have systematic procedures available for manipulating and evaluating such expressions and performing such computations.

In the case of arithmetic, it is elementary algebra that develops the fundamental properties of arithmetic operations and relations.

For example, the operations of addition and multiplication obey familiar laws such as associativity, commutativity and distributivity, while, the "less than or equal" relation satisfies such laws as reflexivity, antisymmetry and transitivity. These laws provide tools which facilitate computation, as well as describe the fundamental nature of numbers, their operations and relations.

The algebra of sets is the set-theoretic analogue of the algebra of numbers. It is the algebra of the set-theoretic operations of union, intersection and complementation, and the relations of equality and inclusion.

THE FUNDAMENTAL LAWS OF SET ALGEBRA

The binary operations of set union and intersection satisfy many identities. Several of these identities or "laws" have well established names. Three pairs of laws, are stated, without proof, in the following proposition.

PROPOSITION 1: For any sets *A*, *B*, and *C*, the following identities hold:

Commutative laws:

- $A \cup B = B \cup A$
- $A \cap B = B \cap A$

Associative laws:

- $(A \cup B) \cup C = A \cup (B \cup C)$
- $(A \cap B) \cap C = A \cap (B \cap C)$

Distributive laws:

- $A \cup (B \cup C) = (A \cup B) \cap (A \cup C)$
- $A \cap (B \cup C) = (A \cup B) \cap (A \cup C)$

Notice that the analogy between unions and intersections of sets, and addition and multiplication of numbers, is quite striking. Like addition and multiplication, the operations of union and intersection are commutative and associative, and intersection *distributes* over unions. However, unlike addition and multiplication, union also *distributes* over intersection.

The next proposition, states two additional pairs of laws involving three specials sets: the empty set, the universal set and the complement of a set.

PROPOSITION 2: For any subset *A* of universal set U, where Ø is the empty set, the following identities hold:

Identity laws:

- $A \cup \varnothing = A$
- $A \cap U = A$

Complement laws:

- $A \cup A^C = U$
- $A \cup A^C = \varnothing$

The identity laws say that, just like 0 and 1 for addition and multiplication, Ø and U are the identity elements for union and intersection, respectively.

Unlike addition and multiplication, union and intersection do not have inverse elements. However the complement laws give the fundamental properties of the somewhat inverse-like unary operation of set complementation. The preceding five pairs of laws: the commutative, associative, distributive, identity and complement laws, can be said to encompass all of set algebra, in the sense that every valid proposition in the algebra of sets can be derived from them.

SOME ONTOLOGY

A set is pure if all of its members are sets, all members of its members are sets, and so on. For example, the set containing only the empty set is a nonempty pure set. When doing set theory, it is common to restrict attention to the pure sets, and many systems of axiomatic set theory are designed to axiomatize the pure sets only. A key idea in set theory is the von Neumann universe of pure sets. Sets in this universe are arranged in a cumulative hierarchy, based on how deeply their members, members of members, etc. are nested. Each set is assigned an ordinal number α in this hierarchy, known as its rank. A set is assigned a rank by transfinite recursion: if the least upper bound on the ranks of the members of a set X is α then X is assigned rank $\alpha+1$. Also, for each ordinal α, the set V_α contains all sets assigned a rank less than α.

AXIOMATIC SET THEORY

Elementary set theory can be studied informally and intuitively, and so can be taught in primary schools using, Venn diagrams. The intuitive approach silently assumes that all objects in the universe of discourse satisfying any defining condition form a set. This assumption gives rise to antinomies, the simplest and best known of which being Russell's paradox. Axiomatic set theory was originally devised to rid set theory of such antinomies.

The most widely studied systems of axiomatic set theory imply that all sets form a cumulative hierarchy. Such systems come in two Flavours, those whose ontology consists of:

- *Sets alone*: This includes the most common axiomatic

set theory, Zermelo–Fraenkel set theory (ZFC), which includes the axiom of choice. Fragments of ZFC include:

- Zermelo set theory, which replaces the axiom schema of replacement with that of separation;
- General set theory, a small fragment of Zermelo set theory sufficient for the Peano axioms and finite sets;
- Kripke-Platek set theory, which omits the axioms of infinity, powerset, and choice, and weakens the axiom schemata of separation and replacement.

- *Sets and proper classes*: This includes Von Neumann-Bernays-Gödel set theory, which has the same strength as ZFC for theorems about sets alone, and Morse-Kelley set theory, which is stronger than ZFC.

The systems NFU and NF, though having their origin in type theory, are not based on a cumulative hierarchy. NF and NFU include a "set of everything," relative to which every set has a complement. Here urelements matter, because NF, but not NFU, produces sets for which the axiom of choice does not hold. Systems of constructive set theory, such as CST, CZF, and IZF, embed their set axioms in intuitionistic logic instead of first order logic. Yet other systems accept standard first order logic but feature a nonstandard membership relation.

These include rough set theory and fuzzy set theory, in which the value of an atomic formula embodying the membership relation is not simply True and False. The Boolean-valued models of ZFC are a related subject.

APPLICATIONS

Nearly all mathematical concepts are now defined formally in terms of sets and set theoretic concepts. For example, mathematical structures as diverse as graphs, manifolds, rings, and vector spaces are all defined as sets having various properties. Equivalence and order relations are ubiquitous in mathematics, and the theory of relations is entirely grounded in set theory. Set theory is also a promising foundational system for much of mathematics. Since the publication of the first volume of *Principia Mathematica,* it has been claimed that most or even all mathematical theorems can be derived using an aptly designed set of axioms for set theory, augmented with many definitions, using first or second order logic. For example, properties of the natural and real numbers can be derived within set theory, as each number system can be identified with a set of equivalence classes under a suitable equivalence relation whose field is some infinite set.

Set theory as a foundation for mathematical analysis, topology, abstract algebra, and discrete mathematics is likewise uncontroversial; mathematicians accept that theorems in these areas can be derived from the relevant definitions and the axioms of set theory. Few full derivations of complex mathematical theorems from set theory have been formally verified, however, because such formal derivations are often much longer than the natural language proofs mathematicians commonly present. One verification project, Metamath, includes derivations of more than 10,000 theorems starting from the ZFC axioms and using first order logic.

AREAS OF STUDY

Set theory is a major area of research in mathematics, with many interrelated subfields.

Combinatorial Set Theory

Combinatorial set theory concerns extensions of finite combinatorics to infinite sets. This includes the study of cardinal arithmetic and the study of extensions of Ramsey's theorem such as the Erdos-Rado theorem.

Descriptive Set Theory

Descriptive Set Theory traces its origins to the theory of integration by Henri Lebesgue at the beginning of 20th century. Investigations into Borel sets of real numbers led to the theory of *projective* sets, and more generally, the theory of definable sets of real numbers. Following Gödel's work, it became apparent that many natural questions in Descriptive Set Theory are undecidable in axiomatic set theory. This was further confirmed by a proliferation of independence results following Cohen's invention of the forcing method. Modern Descriptive Set Theory revolves mostly around the powerful method using infinite games. The branch of Descriptive Set Theory known as *Determinateness,* developed by D. A. Martin, Robert Solovay and others, brought together methods of, among others, Recursion Theory and Large Cardinal Theory and has been very successful in describing the structure of definable sets. More importantly, Descriptive Set Theory provides strong evidence for the large cardinal axioms.

Fuzzy Sets Theory

Fuzzy Set Theory was formalised by Lofti Zadeh at the University of California in 1965. What Zadeh proposed is very much a paradigm shift that first gained acceptance in the Far East and its successful application has ensured its adoption around the world. A paradigm is a set of rules and regulations which defines boundaries and tells us what to do to be successful in solving problems within these boundaries.

For example the use of transistors instead of vacuum tubes is a paradigm shift - likewise the development of Fuzzy Set Theory from conventional bivalent set theory is a paradigm shift. Bivalent Set Theory can be somewhat limiting if we wish to describe a 'humanistic' problem mathematically. For example, Fig bivalent sets to characterise the temperature of a room.

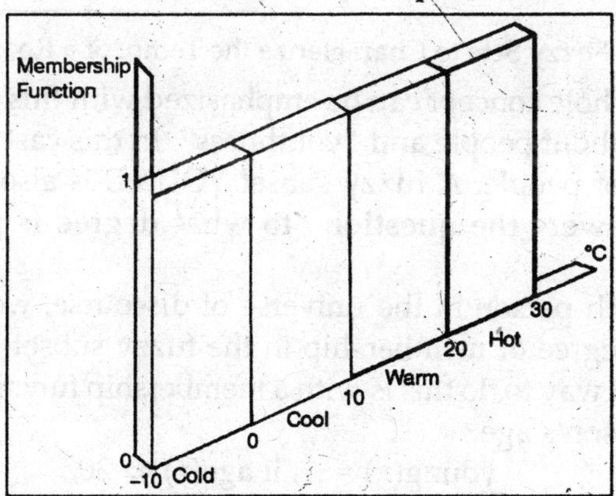

Fig. Bivalent Sets to Characterize the Temp. of a Room.

The most obvious limiting feature of bivalent sets that can be seen clearly from the diagram is that they are mutually exclusive - it is not possible to have membership of more than one set.

Clearly, it is not accurate to define a transiton from a quantity such as 'warm' to 'hot' by the application of one degree Fahrenheit of heat. In the real world a smooth drift from warm to hot would occur. This natural phenomenon can be described more accurately by Fuzzy Set Theory. Fig. shows how fuzzy sets quantifying the same information can describe this natural drift.

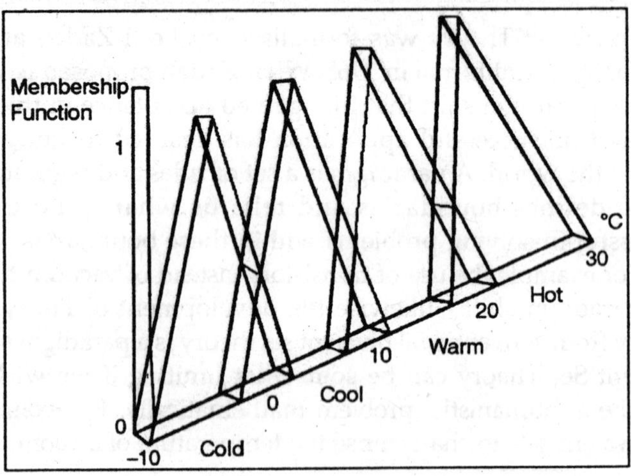

Fig. Fuzzy Sets to Characterize the Temp. of a Room.

The whole concept can be emphasized with this example. Let's talk about people and "youthness". In this case the set S is the set of people. A fuzzy subset YOUNG is also defined, which answers the question "to what degree is person x young?"

To each person in the universe of discourse, we have to assign a degree of membership in the fuzzy subset YOUNG. The easiest way to do this is with a membership function based on the person's age.

$$
\text{young}(x) = \{1, \text{ if age}(x) \Leftarrow 20, \\ (30-\text{age}(x))/10, \text{ if } 20 < \text{age}(x) \Leftarrow 30, \\ 0, \text{ if age}(x) > 30\}
$$

A graph of this looks like:

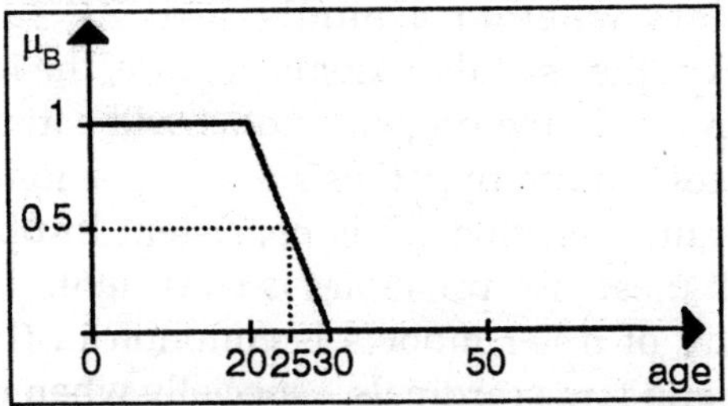

Given this definition, here are some example values:

Person	**Age**	**degree of youth**
Johan	10	1.00
Edwin	21	0.90
Parthiban	25	0.50
Arosha	26	0.40
Chin Wei	28	0.20
Rajkumar	83	0.00

Note: Membership functions almost never have as simple a shape as age(x). They will at least tend to be triangles pointing up, and they can be much more complex than that. Furthermore, membership functions so far is discussed as if they always are based on a single criterion, but this isn't always the case, although it is the most common case.

One could, for example, want to have the membership function for YOUNG depend on both a person's age and their height. This is perfectly legitimate, and occasionally used in practice. It's referred to as a two-dimensional membership function. It's also possible to have even more criteria, or to have the membership function depend on elements from two completely different universes of discourse.

Inner Model Theory

An inner model of Zermelo–Fraenkel set theory (ZF) is a transitive class that includes all the ordinals and satisfies all the axioms of ZF. The canonical example is the constructible universe *L* developed by Gödel. The study of inner models of extensions of ZF is of interest in set theory because it can be used to prove consistency results. For example, it can be shown that regardless whether a model *V* of ZF satisfies the continuum hypothesis or the axiom of choice, the inner model *L* constructed inside the original model will satisfy both the generalized continuum hypothesis and the axiom of choice. Thus the assumption that ZF is consistent implies that ZF together with these two principles is consistent.

The study of inner models is common in the study of determinacy and large cardinals, especially when considering axioms such as the axiom of determinacy that contradict the axiom of choice. Even if a fixed model of set theory satisfies the axiom of choice, it is possible for an inner model to fail to satisfy the axiom of choice. For example, the existence of sufficiently large cardinals implies that there is an inner model satisfying the axiom of determinacy.

Large Cardinals

A large cardinal is a cardinal number with an extra property. Many such properties are studied, including inaccessible cardinals, measurable cardinals, and many more. These properties typically imply the cardinal number must be very large, with the existence of a cardinal with the specified property unprovable in Zermelo-Fraenkel set theory.

Determinacy

Determinacy refers to the fact that, under appropriate assumptions, certain two-player games of perfect information are determined from the start in the sense that one player must have a winning strategy. The existence of these strategies has important consequences in descriptive set theory, as the assumption that a broader class of games is determined often implies that a broader class of sets will have a topological property.

The axiom of determinacy (AD) is an important object of study; although incompatible with the axiom of choice, AD implies that all subsets of the real line are well behaved. AD can be used to prove that the Wadge degrees have an elegant structure.

Forcing

Paul Cohen invented the method of forcing while searching for a model of ZFC in which the axiom of choice or the continuum hypothesis fails. Forcing adjoins to some given model of set theory additional sets in order to create a larger model with properties determined by the construction and the original model. For example, Cohen's construction adjoins additional subsets of the natural numbers without changing any of the cardinal numbers of the original model.

Forcing is also one of two methods for proving relative consistency by finitistic methods, the other method being Boolean-valued models.

Cardinal Invariants

A cardinal invariant is a property of the real line measured by a cardinal number. For example, a well-studied invariant is the smallest cardinality of a collection of meagre sets of reals whose union is the entire real line. These are invariants in the sense that any two isomorphic models of set theory must give the same cardinal for each invariant. Many cardinal invariants have been studied, and the relationships between them are often complex and related to axioms of set theory.

Set-theoretic Topology

Set-theoretic topology studies questions of general topology that are set-theoretic in nature or that require advanced methods of set theory for their solution. Many of these theorems are independent of ZFC, requiring stronger axioms for their proof. A famous problem is the normal Moore space question, a question in general topology that was the subject of intense research. The answer to the normal Moore space question was eventually proved to be independent of ZFC.

OBJECTIONS TO SET THEORY AS A FOUNDATION FOR MATHEMATICS

From set theory's inception, some mathematicians objected to it as a foundation for mathematics, arguing, for example, that it is just a game which includes elements of fantasy. The most common objection to set theory, one Kronecker voiced in set theory's earliest years, starts from the constructivist view that mathematics is loosely related to computation. If this view is granted, then the treatment of infinite sets, both in naive and in axiomatic set theory, introduces into mathematics methods and objects that are not computable even in principle.

Ludwig Wittgenstein questioned the way Zermelo-Fraenkel set theory handled infinities. Wittgenstein's views about the foundations of mathematics were later criticised by Paul Bernays, and closely investigated by Crispin Wright, among others. Category theorists have proposed topos theory as an alternative to traditional axiomatic set theory. Topos theory can interpret various alternatives to that theory, such as constructivism, finite set theory, and computable set theory.

COMPOUND INTEREST

Compound interest is interest calculated on the principal amount invested, which is then added to the principal amount, and compounded again. Compound interest can be earned daily, weekly, monthly or yearly. Generally the more times an amount is compounded, the more money you can make.

As long as you leave an interest earning account alone, by not removing money from it, you begin making more money on your investment because the money you earn is added back to the principle amount. It's a simple fact that more money earning interest makes you more money. Each time interest is compounded, the money earned gets added to the total.

If you were raising two rabbits, you might view a similar thing. If the bunnies produced a litter, and you kept all those bunnies, then you might have possibly eight rabbits. The original bunnies would keep on breeding, as would the new litter, and you'd end up with more rabbits then you knew what to do with. Compound interest won't be quite that dramatic, unless you're investing huge sums of money.

The important parallel is that the first pair of bunnies and their offspring now combine together to produce yet more bunnies, and as combined, they will produce a great deal more than if they were sold off and separated. Most investment firms, banks, and the like, will state how often your interest is compounded. In some cases, your investment doesn't compound, but earns what is called simple interest. This means you only make money on the amount you initially invested, and the profits are not reinvested to make you more money.

You can figure out exactly how much an investment will be worth in a few years if you have a scientific calculator handy. You also need to know the initial investment amount (principal or p), the rate of interest, (r), the number of years you plan to allow the investment to sit (years or y) and the number of times per year you investment will compound (t).

That only a portion of the interest would be earned each month, so the interest amount would have to be divided by the total times interest gets compounded each year (t).

The formula is as follows:

Total value = p(1 + r/t)ty

Putting this to work, in dollar amounts, you might invest $10,000 US Dollars (USD) in a savings account that earns 5% interest per year and is compounded monthly. If you leave that money alone for five years, you could figure out exactly how much money you'd make in that time period, and the value of your account at the end of four years.

The equation would look like this:

$$10{,}000(1 + .05/12)^{12 \times 5} = \$12{,}833.59$$

you only earned simple interest, at even 5.5% per year, you wouldn't make that much money.

Note the following:

$$10{,}000(1 + .055 \times 5) = \$12{,}750.00$$

One reason to understand compound interest is because some accounts that earn simple interest offer a higher yearly interest rate. Yet if your investment is long term, you may make more money with a lower interest rate that compounds your interest.

On the other hand, if you know you'll be removing the money after a year or two, a higher interest rate that is not compounded may be a better investment, than an account with compound interest at a lower rate.

Also, don't be daunted by these formulas if you are calculating interest. If you have access to the Internet, you can find hundreds of sites that offer compound interest calculators and most of them are very easy to use.

MATHEMATICS OF INTEREST RATES

Simplified Calculation

Formulae are presented in greater detail at time value of money. In the formula, i is the effective interest rate per period. FV and PV represent the future and present value of a sum. n represents the number of periods.

These are the most basic formula:

$$FV = PV(1+i)^n$$

The calculates the future value (FV) of an investment's present value (PV) accruing at a fixed interest rate (i) for n periods.

$$PV = \frac{FV}{(1+i)^n}$$

The calculates what present value (PV) would be needed to produce a certain future value (FV) if interest (i) accrues for n periods.

$$i = \left(\frac{FV}{PV}\right)^{\frac{1}{n}} - 1$$

The calculates the compound interest rate achieved if an initial investment of PV returns a value of FV after n accrual periods.

$$n = \frac{\log(FV) - \log(PV)}{\log(1+i)}$$

The formula calculates the number of periods required to get FV given the PV and the interest rate (i). The log function can be in any base, e.g. natural log (ln)

Compound

Formula for calculating compound interest:

$$A = P\left(1+\frac{r}{n}\right)^{nt}$$

Where,

- P = Principal amount (initial investment)
- r = Annual nominal interest rate (as a decimal)
- n = Number of times the interest is compounded per year
- t = Number of years
- A = Amount after time t

Example usage: An amount of $1500.00 is deposited in a bank paying an annual interest rate of 4.3%, compounded quarterly. Find the balance after 6 years.

A. Using the formula, with P = 1500, r = 4.3/100 = 0.043, n = 4, and t = 6:

$$A = 1500\left(1 + \frac{0.043}{4}\right)^{4\times 6} = 1938.84$$

So, the balance after 6 years is approximately $1,938.op.

Periodic Compounding

The amount function for compound interest is an exponential function in terms of time.

$$A(t) = A_0\left(1 + \frac{r}{n}\right)^{nt}$$

- *t* = Total time in years
- *n* = Number of compounding periods per year (note that the total number of compounding periods is $n \cdot t$)
- *r* = Nominal annual interest rate expressed as a decimal. e.g.: 6% = 0.06

As *n* increases, the rate approaches an upper limit of e^r " 1. This rate is called *continuous compounding.*

Since the principal *A(0)* is simply a coefficient, it is often dropped for simplicity, and the resulting accumulation function is used in interest theory instead. Accumulation functions for simple and compound interest are:

$$a(t) = 1+tr$$

$$a(t) = \left(1 + \frac{r}{n}\right)^{nt}$$

Note: $A(t)$ is the amount function and $a(t)$ is the accumulation function.

Continuous Compounding

Continuous compounding can be thought as making the compounding period infinitesimally small; therefore achieved by taking the limit of n to infinity. One should consult definitions of the exponential function for the mathematical proof of this limit.

$$P_1 = P_0 e^{rt} \text{ take t as 1 so}$$

$$ln(P_1 / P_0) = r = ln(1 + R)$$

where R is simple return and r is called log return because it is the logarithm of normal return.

$$a(t) = \lim_{n \to \infty} \left(1 + \frac{r}{n}\right)^{nt}$$

$$a(t) = e^{rt}$$

The amount function is simply,

$$A(t) = A_0 e^{rt}$$

The interest rate expressed as a continuously compounded rate is called the force of interest. The annual force of interest is simply 12 times the monthly force of interest.

The effective interest rate per year is,

$i = e^r - 1$

Using this i the amount function can be written as:

$$A(t) = A_0(1 + i)^t$$

or

$$A = P(1 + i)^t$$

Force of Interest

In mathematics, the accumulation functions are often expressed in terms of *e*, the base of the natural logarithm. This facilitates the use of calculus methods in manipulation of interest formulae. For any continuously differentiable accumulation function *a(t)* the force of interest, or more generally the logarithmic or continuously compounded return is a function of time defined as follows:

$$\delta_t = \frac{a'(t)}{a(t)}$$

which is the rate of change with time of the natural logarithm of the accumulation function.

Conversely:

$$a(n) = e^{\int_0^n \delta t \, dt} \text{ (since } a(0) = 1)$$

When the formula is written in differential equation format, the force of interest is simply the coefficient of amount of change.

$$da(t) = \delta_t a(t) dt$$

For compound interest with a constant annual interest rate *r* the force of interest is a constant, and the accumulation function of compounding interest in terms of force of interest is a simple power of e:

$$\delta - \text{In}(1+r)$$

$$a(t) = e^{t\delta}$$

The force of interest is less than the annual effective interest rate, but more than the annual effective discount rate. It is the reciprocal of the e-folding time.

Compounding Basis

To convert an interest rate from one compounding basis to another compounding basis, the following formula applies:

$$r_2 = \left[\left(1+\frac{r_1}{n_1}\right)^{\frac{n1}{n2}} - 1\right] n_2$$

where r_1 is the stated interest rate with compounding frequency n_1 and r_2 is the stated interest rate with compounding frequency n_2.

When interest is continuously compounded:

$$R = n\text{In}(1 + r/n)$$

where *R* is the interest rate on a continuous compounding basis and *r* is the stated interest rate with a compounding frequency *n*.

U.S. Monthly Mortgage Payments

The interest on U.S. mortgages is compounded monthly. The formula for payments is found from the following argument.

Notation

I = Note percentage rate
i = Monthly percentage rate
= I/12 (so that the APR = (1+i)^12)
T = Term in years
Y = IT
X = 1/2 I T = 1/2 Y
n = 12 T = term in months
L = Principal or amount of loan
P = monthly payment

Exact Formula for P

If the term were only one month then clearly,

$$(1 + i)L = P \text{ so that } L = \frac{P}{1+i}.$$

If the term were two months then,

$$(1 + i)((1 + i)L \text{ “ } P) = P \text{ so that } L = \frac{P}{1+i} + \frac{P}{(1+i)^2}.$$

For a term of n months then,

$$L = P\sum_{j=1}^{n} \frac{1}{(1+i)^j}.$$

This can be simplified by noting that,

$$(1+i)L = P\sum_{j=0}^{n-1} \frac{1}{(1+i)^j}$$

and taking the difference:

$$(1+i)L - L = iL = P\left(1 - \frac{1}{(1+i)^n}\right)$$

so that

$$P = \frac{Li}{1 - \frac{1}{(1+i)n}} = \frac{Li}{1 - e^{-n\ln(1+i)}}$$

This formula for the monthly payment on a U.S. mortgage is exact and is what banks use.

Approximate Formula for P

A formula that is accurate to within a few per cent can be found by noting that for typical U.S. note rates (*I* < 8% and terms (T=10-30 years), that the monthly note rate is small compared to 1: *i* < < 1 so that the $\text{In}(1+i) \approx i$ which yields a simplification so that,

$$P \approx \frac{Li}{1-e^{-ni}} = \frac{L}{n}\frac{ni}{1-e^{-ni}}$$

which suggests defining a auxiliary variables,

$$Y \equiv ni = TI$$

$$P_0 \equiv \frac{L}{n}.$$

P_0 is the monthly payment required for a zero interest loan paid off in *n* installements. In terms of these variables the approximation can be written,

$$P \approx P_0 \frac{Y}{1-e^{-Y}}$$

The function,

$$f(Y) \equiv \frac{Y}{1-e^{-Y}} - \frac{Y}{2}$$

is even: $f(Y) = f(-Y)$ implying that it can be expanded in even powers of *Y*.

It follows immediately that,

$$\frac{Y}{1-e^{-Y}}$$

can be expanded in even powers of *Y* plus the single term: $Y/2$

It will prove convenient then to define,

$$X = \frac{1}{2}Y = \frac{1}{2}IT$$

so that

$$P \approx P_0 \frac{2X}{1-e^{-2X}}$$

which can be expanded:

$$P \approx P_0 \left(1 + X + \frac{X_2}{3} - \frac{1}{45}X^4 + \ldots\right)$$

where the ellipses indicate terms that are higher order in even powers of X. The expansion,

$$P \approx P_0\left(1 + X + \frac{X_2}{3}\right)$$

is valid to better than 1% provided $X \leq 1$.

Example

For a mortgage with a term of 30 years and a note rate of 4.5% we find:

$T = 30$

$I = .045$

$$X = \frac{1}{2}IT = \frac{1}{2} \times .045 \times 30 = .675$$

which suggests that the approximation

$$P \approx P_0\left(1 + X + \frac{1}{3}X^2\right)$$

is accurate to better than one per cent for typical U.S. mortgage terms in January 2009. The formula becomes less accurate for higher rates and longer terms.

For a 30-year term on a loan of \$120,000 and a 4.5% note rate we find:

$$L = 120000$$

$$P_0 = \frac{\$120{,}000}{360} = \$333.33$$

so that,

$$P \approx P_0 = \left(1 + X + \frac{1}{3}X^2\right) = \$333.33\left(1 + .675 + .675^2 / 3\right)\$608.96$$

The exact payment amount is P = \$608.02 so the approximation is an overestimate of about a sixth of a per cent.

Other Approximations

The approximate formula,

$$P \approx P_0 = \frac{Y}{1 - e^{-Y}}$$

yields $P_0 \approx \$607.47$ which is a slight underestimate of the exact result. This underestimate results from the approximation

$$\text{In}(1 + i) \approx i.$$

Keeping the next correction in the expansion of,

$$\ln(1+i) \approx i - i^2/2 + ...$$

results in an approximate formula

$$P \approx P_0 \frac{Y}{1-e^{-Y}e^{iY/2}} = \$608.018$$

which is off by two tenths of a cent.

$$P \approx P_0 \left(1 + X + \frac{1}{3}X^2\right)$$

The simplest approximation discussed is good to within better than a per cent for typical US mortgages in early 2009. The approximation,

$$P \approx P_0(1+X)$$

is an underestimate of around 10% for such mortgages.

DEPRECIATION AND ANNUITIES

DEPRECIATION

Depreciation is a term used in accounting, economics and finance to spread the cost of an asset over the span of several years.

In common, depreciation is the reduction in the value of an asset used for business purposes during certain amount of time due to usage, passage of time, wear and tear, technological outdating or obsolescence, depletion, inadequacy, rot, rust, decay or other such factors.

In accounting, however, depreciation is a term used to describe any method of attributing the historical or purchase cost of an asset across its useful life, roughly corresponding to normal wear and tear. It is of most use when dealing with assets of a short, fixed service life, and which is an example of applying the matching principle per generally accepted accounting principles. Depreciation in accounting is often mistakenly seen as a basis for recognizing impairment of an asset, but unexpected changes in value, where seen as significant enough to account for, are handled through write-downs or similar techniques which adjust the book value of the asset to reflect its current value.

The use of depreciation affects the financial statements and in some countries the taxes of companies and individuals. The recording of depreciation will cause an expense to be recognized, thereby lowering stated profits on the income statement, while the net value of the asset will decline on the balance sheet. Depreciation reported for accounting and tax purposes may differ substantially.

Depreciation and its related concept, amortization are non-cash expenses. Neither depreciation nor amortization will directly affect the cash flow of a company, as both are accounting representations of expenses attributable to a given period. In accounting statements, depreciation may neither figure in the cash flow statement, nor be "added back" to net income to derive the operating cash flow. Depreciation recognized for tax purposes will, however, affect the cash flow of the company, as tax depreciation will reduce taxable profits; there is generally no requirement that treatment of depreciation for tax and accounting purposes be identical. Where depreciation is shown on accounting statements, the figure usually does not match the depreciation for tax purposes.

Because of its non-standardized derivation, depreciation is a key component of EBITDA, a metric used to gauge the worth of a company independent of tax-jurisdiction effects and capitalization structure.

Salvage value is the estimated value of an asset at the end of its useful life. In accounting, the salvage value of an asset is its remaining value after depreciation. This is also known as residual value or scrap value. It is the net cash inflow that occurs when the asset is liquefied at the end of its life. Salvage value can be negative if the residual asset requires special treatment to terminate—for example, used nuclear materials or CRT's containing lead.

In economics, depreciation is the decrease in the economic value of the capital stock of a firm, nation or other entity, either through physical depreciation, obsolescence or changes in the demand for the services of the capital in question. If capital stock is C_0 at the beginning of a period, investment is I and depreciation D, the capital stock at the end of the period, C_1, is $C_0 + I - D$.

Accounting

A company needs to report depreciation accurately in its financial statements in order to achieve two main objectives:

- Matching its expenses with the income generated by means of those expenses, and
- Ensuring that the asset values in the balance sheet are not overstated.

Depreciation is an attempt to write-off the cost of Non Current Asset over its useful life. The word write-off means to turn it into an expense. For example, an entity may depreciate its equipment by 15% per year. This rate should be reasonable in aggregate and consistently employed. However, there is no expectation that each individual item declines in value by the same amount, primarily because the recognition of depreciation is based upon the allocation of historical costs and not current market prices.

Accounting standards bodies have detailed rules on which methods of depreciation are acceptable, and auditors should express a view if they believe the assumptions underlying the estimates do not give a true and fair view.

Recording Depreciation

For historical cost purposes, assets are recorded on the balance sheet at their original cost; this is called the historical cost. Historical cost minus all depreciation expenses recognized on the asset since purchase is called the book value. Depreciation is not taken out of these assets directly.

It is instead recorded in a contra asset account: an asset account with a normal credit balance, typically called "accumulated depreciation". Balancing an asset account with its corresponding accumulated depreciation account will result in the net book value. The net book value will never fall below the salvage value, meaning that once an asset is fully depreciated, no further expenses will be taken during its life. Salvage value is the estimated value of the asset at the end of its useful life. In this way, total depreciation for an asset will never exceed the estimated total cash outlay for the asset. The exception to this is in many price-regulated industries where salvage is estimated net of the cost of physically removing the asset from service.

If the expected cost of removal exceeds the expected raw salvage, then the net of the two may be negative. In this case, the depreciation recorded on the regulated books may exceed the depreciable basis. Companies have no obligation to dispose of depreciated assets, of course, and many fully depreciated assets continue to generate income. Recording a depreciation expense will involve a credit to an accumulated depreciation account.

The corresponding debit will involve either an expense account or an asset account that represents a future expense, such as work in progress. Depreciation is recorded as an adjusting journal entry. A write-down is a form of depreciation that involves a partial write off. Part of the value of the asset is removed from the balance sheet. The reason may be that the book value of the fixed asset has diverged from the market value and causes the company a loss. An example of this would be a revaluation of goodwill on an acquisition that went bad.

Methods of Depreciation

There are several methods for calculating depreciation, generally based on either the passage of time or the level of activity of the asset.

Straight-line Depreciation

Straight-line depreciation is the simplest and most-often-used technique, in which the company estimates the salvage value of the asset at the end of the period during which it will be used to generate revenues and will expense a portion of original cost in equal increments over that period.

The salvage value is an estimate of the value of the asset at the time it will be sold or disposed of; it may be zero or even negative. Salvage value is also known as scrap value or residual value.

Straight-line method:

$$\text{annual depreciation expense} = \frac{\text{cost of fixed asset} - \text{residual value}}{\text{useful life of asset (years)}}$$

For example, a vehicle that depreciates over 5 years, is purchased at a cost of US$17,000, and will have a salvage value of US$2000, will depreciate at US$3,000 per year: ($17,000 " $2,000)/ 5 years = $3,000 annual straight-line depreciation expense. In other words, it is the depreciable cost of the asset divided by the number of years of its useful life.

This table emphasizes the straight-line method of depreciation. Book value at the beginning of the first year of depreciation is the original cost of the asset.

At any time book value equals original cost minus accumulated depreciation. book value = original cost " accumulated depreciation Book value at the end of year becomes book value at the beginning of next year. The asset is depreciated until the book value equals scrap value.

Book value beginning of year	Depreciation expense	Accumulatedde preciation	Book value atend of year at
$17,000 (original cost)	$3,000	$3,000	$14,000
$14,000	$3,000	$6,000	$11,000
$11,000	$3,000	$9,000	$8,000
$8,000	$3,000	$12,000	$5,000
$5,000	$3,000	$15,000	$2,000 (scrap value)

If the vehicle were to be sold and the sales price exceeded the depreciated value then the excess would be considered a gain and subject to depreciation recapture. In addition, this gain above the depreciated value would be recognized as ordinary income by the tax office. If the sales price is ever less than the book value, the resulting capital loss is tax deductible. If the sale price were ever more than the original book value, then the gain above the original book value is recognized as a capital gain.

If a company chooses to depreciate an asset at a different rate from that used by the tax office then this generates a timing difference in the income statement due to the difference between the taxation department's and company's view of the profit.

Declining-balance Method

Depreciation methods that provide for a higher depreciation charge in the first year of an asset's life and gradually decreasing charges in subsequent years are called accelerated depreciation methods. This may be a more realistic reflection of an asset's actual expected benefit from the use of the asset: many assets are most useful when they are new. One popular accelerated method is the declining-balance method. Under this method the book value is multiplied by a fixed rate.

annual depreciation = depreciation rate * book value at beginning of year

The most common rate used is double the straight-line rate. For this reason, this technique is referred to as the double-declining-balance method. To emphasize, suppose a business has an asset with $1,000 original cost, $100 salvage value, and 5 years useful life. First, calculate straight-line depreciation rate. Since the asset has 5 years useful life, the straight-line depreciation rate equals (100%/ 5) 20% per year. With double-declining-balance method, as the name suggests, double that rate, or 40% depreciation rate is used. The table emphasizes the double-declining-balance method of depreciation.

Book value at beginning of year	Depreciation rate	Depreciation expense	Accumulated depreciation	Book value at end of year
$1,000 (original cost)	40%	$400	$400	$600
$600	40%	$240	$640	$360
$360	40%	$144	$784	$216
$216	40%	$86.40	$870.40	$129.60
$129.60	$129.60 - $100	$29.60	$900	$100 (scrap value)

When using the double-declining-balance method, the salvage value is not considered in determining the annual depreciation, but the book value of the asset being depreciated is never brought below its salvage value, regardless of the method used.

The process continues until the salvage value or the end of the asset's useful life, is reached. In the last year of depreciation a subtraction might be needed in order to prevent book value from falling below estimated Scrap Value.

Since double-declining-balance depreciation does not always depreciate an asset fully by its end of life, some methods also compute a straight-line depreciation each year, and apply the greater of the two.

This has the effect of converting from declining-balance depreciation to straight-line depreciation at a midpoint in the asset's life. It is possible to find a rate that would allow for full depreciation by its end of life with the formula:

$$\text{depreciation rate} = 1 - \sqrt[N]{\frac{\text{residual value}}{\text{cost of fixed asset}}},$$

where N is the estimated life of the asset.

Activity Depreciation

Activity depreciation methods are not based on time, but on a level of activity. This could be miles driven for a vehicle, or a cycle count for a machine. When the asset is acquired, its life is estimated in terms of this level of activity. Assume the vehicle above is estimated to go 50,000 miles in its lifetime.

The per-mile depreciation rate is calculated as: ($17,000 cost - $2,000 salvage)/ 50,000 miles = $0.30 per mile. Each year, the depreciation expense is then calculated by multiplying the rate by the actual activity level.

Sum-of-years' Digits Method

Sum-of-years' digits is a depreciation method that results in a more accelerated write-off than straight line, but less than declining-balance method. Under this method annual depreciation is determined by multiplying the Depreciable Cost by a schedule of fractions.

depreciable cost = original cost – salvage value

book value = original cost – accumulated depreciation

Example: If an asset has original cost of $1000, a useful life of 5 years and a salvage value of $100, compute its depreciation schedule. First, determine years' digits. Since the asset has useful life of 5 years, the years' digits are: 5, 4, 3, 2, and 1.

Next, calculate the sum of the digits. 5+4+3+2+1=15

The sum of the digits can also be determined by using

the formula $(n^2+n)/2$ where n is equal to the useful life of the asset. The example would be shown as $(5^2+5)/2=15$

Depreciation rates are as follows: 5/15 for the 1st year, 4/15 for the 2nd year, 3/15 for the 3rd year, 2/15 for the 4th year, and 1/15 for the 5th year.

Book value at beginning of year	Total depreciable cost	Depreciation rate	Depreciation expense	Accumulated depreciation	Book value at end of year
$1,000 (original cost)	$900	5/15	$300 ($900 * 5/15)	$300	$700
$700	$900	4/15	$240 ($900 * 4/15)	$540	$460
$460	$900	3/15	$180 ($900 * 3/15)	$720	$280
$280	$900	2/15	$120 ($900 * 2/15)	$840	$160
$160	$900	1/15	$60 ($900 * 1/15)	$900	$100 (scrap value)

Units-of-production depreciation method

Under the units-of-production method, useful life of the asset is expressed in terms of the total number of units expected to be produced:

annual depreciation expense =

$$\frac{\text{cost of fixed asset} - \text{residual value}}{\text{estimated total production}} \times \text{actual production}$$

Suppose, an asset has original cost $70,000, salvage value $10,000, and is expected to produce 6,000 units.

Depreciation per unit = ($70,000"10,000)/ 6,000 = $10

10 × actual production will give you the depreciation cost of the current year.

The table emphasizes the units-of-production depreciation schedule of the asset.

Book value at beginning of year	Units of production	Depreciation cost per unit	Depreciation expense	Accumulated depreciation	Book value at end of year
$70,000 (original cost)	1,000	$10	$10,000	$10,000	$60,000
$60,000	1,100	$10	$11,000	$21,000	$49,000
$49,000	1,200	$10	$12,000	$33,000	$37,000
$37,000	1,300	$10	$13,000	$46,000	$24,000
$24,000	1,400	$10	$14,000	$60,000	$10,000 (scrap value)

Depreciation stops when book value is equal to the Scrap Value of the asset. In the end the sum of accumulated depreciation and scrap value equals to the original cost.

Units of Time Depreciation

Units of time depreciation is similar to units of production, and is used for depreciation equipment used in mine or natural resource exploration, or cases where the amount the asset is used is not linear year to year.

A simple example can be given for construction companies, where some equipment is used only for some specific purpose. Depending on the number of projects, the equipment will be used and depreciation charged accordingly.

Group Depreciation Method

Group depreciation method is used for depreciating multiple-asset accounts using straight-line-depreciation method. Assets must be similar in nature and have approximately the same useful lives.

Asset	Historical cost	Salvage value	Depreciable cost	Life	Depreciation per year
Computers	$5,500	$500	$5,000	5	$1,000

Composite depreciation method

The composite method is applied to a collection of assets that are not similar, and have different service lives. For example, computers and printers are not similar, but both are part of the office equipment. Depreciation on all assets is determined by using the straight-line-depreciation method.

Asset	Historical cost	Salvage value	Depreciable cost	Life	Depreciation per year
Computers	$5,500	$500	$5,000	5	$1,000
Printers	$1,000	$100	$ 900	3	$ 300
Total	$ 6,500	$600	$5,900	4.5	$1,300

Composite life equals the total depreciable cost divided by the total depreciation per year. $5,900/ $1,300 = 4.5 years.

Composite depreciation rate equals depreciation per year divided by total historical cost. $1,300/ $6,500 = 0.20 = 20%

Depreciation expense equals the composite depreciation rate times the balance in the asset account (historical cost). (0.20 * $6,500) $1,300. Debit depreciation expense and credit accumulated depreciation. When an asset is sold, debit cash for the amount received and credit the asset account for its original cost. Debit the difference between the two to accumulated depreciation. Under the composite method no gain or loss is recognized on the sale of an asset. Theoretically, this makes sense because the gains and losses from assets sold before and after the composite life will average themselves out.

To calculate composite depreciation rate, divide depreciation per year by total historical cost. To calculate depreciation expense, multiply the result by the same total historical cost. The result, not surprisingly, will equal to the total depreciation Per Year again.

Common sense requires depreciation expense to be equal to total depreciation per year, without first dividing and then multiplying total depreciation per year by the same number. Creators of accounting rules sometimes are very creative, as was noted on the discussion forum of accounting coach at

Taxes

When a company spends money for a service or anything else that is short-lived, this expenditure is usually immediately tax deductible in some countries, and the company enjoys an immediate tax benefit.

To be eligible for depreciation, an asset must have two features:

- It has a useful life beyond the taxable year, and
- it wears out, decays, declines in value due to natural causes, or is subject to exhaustion or obsolescence.

Therefore, when a company buys an asset that will last longer than one year, like a computer, car, or building, the company cannot immediately deduct the cost and enjoy an immediate large tax benefit. Instead, the company must depreciate the cost over the useful life of the asset, taking a tax deduction for a part of the cost each year. Eventually the company does get to deduct the full cost of the asset, but this happens over several years.

In the US, the IRS's depreciation schedule for any given class of asset is fixed, and is related to typical durability. A computer may depreciate completely over five years; a nonresidential building, usually 39 years. The maximum allowable useful life under US income tax regulations is 40 years.

Though the IRS does allow a small choice of permutations for depreciation life and acceleration, it does not allow a taxpayer to invent any arbitrary asset life. Other countries have other systems, many simply eliminate all choice altogether.

In these jurisdictions accounting depreciation and tax depreciation are almost always significantly different numbers, as in many instances a form of"accelerated depreciation" can be used for tax purposes to lower net income in a given period. Technically, these are not considered"tax reductions" but tax deferrals: lowering taxable income now by increasing expenses should increase future taxable income at a later date.

Importantly, no depreciation deduction is allowed for inventories or other property held for sale to customers in the ordinary course of business. Land is also not depreciable. However, improvements to land, including landscaping, are usually depreciable.

In the U.S., there are generally five variables that a taxpayer must take into account when computing the correct depreciation deduction:

- The depreciation base,
- The asset's class life,
- The applicable recovery period,
- The applicable depreciation method, and
- The applicable convention.

Economics

Models

In economics, the value of a capital asset may be modeled as the present value of the flow of services the asset will generate in future, appropriately adjusted for uncertainty. Economic depreciation over a given period is the reduction in the remaining value of future services. Under certain circumstances, such as an unanticipated increase in the price of the services generated by an asset or a reduction in the discount rate, its value may increase rather than decline.

Depreciation is then negative. Depreciation can alternatively be measured as the change in the market value of capital over a given period: the market price of the capital at the beginning of the period minus its market price at the end of the period.

Such a method in calculating depreciation differs from other methods, such as straight-line depreciation in that it is included in the calculation of implicit cost, and thus economic profit.

Modeling depreciation of a durable as delivering the same services from purchase until failure, with zero scrap value is referred to as the light bulb model of depreciation, or more colourfully as the one-hoss shay model, after a poem by Oliver Wendell Holmes, Sr., about a carriage which worked perfectly for exactly one hundred years, then fell completely apart in an instant.

National Accounts

In national accounts the decline in the aggregate capital stock arising from the use of fixed assets in production is referred to as consumption of fixed capital. Hence, CFC is equal to the difference between aggregate gross fixed capital formation and net fixed capital formation or between Gross National Product and Net National Product.

Unlike depreciation in business accounting, CFC in national accounts is, in principle, not a method of allocating the costs of past expenditures on fixed assets over subsequent accounting periods. Rather, fixed assets at a given moment in time are valued just as to the remaining benefits to be derived from their use.

ANNUITY

An annuity is an investment that you make, either in a single lump sum or through installments paid over a certain number of years, in return for which you receive back a specific sum every year, every half-year or every month, either for life or for a fixed number of years. After the death of the annuitant, or after the fixed annuity period expires for annuity payments, the invested annuity fund is refunded, perhaps along with a small addition, calculated at that time.

Annuities differ from all the other forms of life insurance discussed so far in one fundamental way - an annuity does not provide any life insurance cover but, instead, offers a guaranteed income either for life or a certain period.

Typically annuities are bought to generate income during one's retired life, which is why they are also called pension plans. Annuity premiums and payments are fixed with reference to the duration of human life. Annuities are an investment, which can offer an income you cannot outlive and provide a solution to one of the biggest financial insecurities of old age; namely, of outliving one's income.

2

Equations

INTRODUCTION

EQUATIONS AND REALITY

There are several different kinds or categories of equation, all of which require a different approach to their solution. Indeed, the fundamental nature of the solution differs significantly for the various categories.

There is also a connection between the physical system and the equations which describe it, in that certain types of equation describe certain situations or phenomena. An understanding of the relationship between the `real' situation and the types of equations is thus essential in developing mathematical models in the form of equations.

CLASSIFICATION OF EQUATIONS AND SYSTEMS

- What kinds of equation are there?
- How do these relate to engineering situations?
- What is the nature of the solution?

WHAT IS AN EQUATION?

An equation is anything with an equals sign in it, i.e.: (Something) = (Something else) There are three major categories of equation.

ALGEBRAIC EQUATIONS

These are most easily identifiable by what they do not

contain, i.e. No derivatives or integrals but any of the normal arithmetic operators or higher algebraic functions.

The following are thus all algebraic equations:

$$x - 3 = 5$$

$$\sin 3x = y \cos x$$

$$a x^2 + b x + c = 0$$

The solution is a numerical value for a single equation, or a set of numbers, as many as there are unknowns and equations, for a set of algebraic equations.

One significant characteristic of an equation is whether it is linear or nonlinear in a particular variable. The equation is linear only if the variable appears to a power of one, and does not appear as the argument of a higher function.

Thus the equation is linear in x. However the second equation is linear in y, and the third is linear in a, b, and c.

ORDINARY DIFFERENTIAL EQUATIONS: O.D.ES

These are readily identifiable as they contain derivatives:

$$\frac{dy}{dt} = \sin t - y$$

$$kA\frac{dT}{dx} = -Q$$

$$\frac{d^2y}{dt^2} + \frac{dy}{dt} - ay = 0$$

NB: These must all be `Straight d' derivatives if the equations are o.d.e.s.

There are two different kinds of variable in this type of equation: the dependent and independent variables. There may only be one independent variable, and this will appear on the bottom line of the derivatives.

The variables which appear on the top line are the dependent variables and there should be the same number of these as there are equations.

The solution is the dependent variable or variables as a function or set of functions in terms of the independent variable.

PARTIAL DIFFERENTIAL EQUATIONS: P.D.E.S

If there is more than one independent variable, then derivatives must normally be written as partial derivatives, giving rise to p.d.e.s with `curly ds', e.g.:

$$a\frac{\partial^2 T}{\partial x^2}+b\frac{\partial T}{\partial t}=c$$

P.d.e.s are hard to solve! The solution is the dependent variables as functions of all the independent variables.

LINEAR EQUATION

A linear equation is an algebraic equation in which each term is either a constant or the product of a constant and (the first power of) a single variable.

Linear equations can have one or more variables. Linear equations occur with great regularity in applied mathematics. While they arise quite naturally when modeling many phenomena, they are particularly useful since many non-linear equations may be reduced to linear equations by assuming that quantities of interest vary to only a small extent from some"background" state.

LINEAR EQUATIONS IN TWO VARIABLES

A common form of a linear equation in the two variables x and y is

$$y = mx + b,$$

where m and b designate constants. The origin of the name"linear" comes from the fact that the set of solutions of such an equation forms a straight line in the plane. In this particular equation, the constant m determines the slope or gradient of that line, and the constant term b determines the point at which the line crosses the y-axis, otherwise known as the y-intercept.

Since terms of a linear equations cannot contain products of distinct or equal variables, nor any power (other than 1) or other function of a variable, equations involving terms such as xy, x^2, $y^{1/3}$, and sin(x) are nonlinear.

Forms for 2D Linear Equations

Linear equations can be rewritten using the laws of elementary algebra into several different forms. These equations are often referred to as the"equations of the straight line". In what follows x, y and t are variables; other letters represent constants (fixed numbers).

General Form

$$Ax + By + C = 0$$

where A and B are not both equal to zero. The equation is usually written so that A ⇐ 0, by convention. The graph of the equation is a straight line, and every straight line can be represented by an equation in the form.

If A is nonzero, then the x-intercept, that is the x-coordinate of the point where the graph crosses the x-axis (y is zero), is –C/A.

If B is nonzero, then the y-intercept, that is the y-coordinate of the point where the graph crosses the y-axis (x is zero), is –C/B, and the slope of the line is –A/B.

Standard Form

$$Ax + By = C,$$

where A, B, and C are integers whose greatest common factor is 1, A and B are not both equal to zero, and A is non-negative (and if A = 0 then B has to be positive).

The standard form can be converted to the general form, but not always to all the other forms if A or B is zero. It is worth noting that, while the term occurs frequently in school-level US textbooks, it makes little mathematical sense since most lines cannot be described by such equations. For instance, the line x + y = Ö2 cannot be described by a linear equation with integer coefficients since Ö2 is irrational.

Slope-intercept Form

$$y = mx + b$$

where m is the slope of the line and b is the y-intercept, which is the y-coordinate of the point where the line crosses the y axis. This can be seen by letting x = 0, which immediately gives y = b. Vertical lines, having undefined slope, cannot be represented by this form.

Point-slope form

$$y - y_1 = m(x - x_1)$$

where m is the slope of the line and (x1,y1) is any point on the line. The point-slope and slope-intercept forms are easily interchangeable.

The point-slope form expresses the fact that the difference in the y coordinate between two points on a line (that is, $y - y_1$) is proportional to the difference in the x coordinate (that is, $x - x_1$). The proportionality constant is m (the slope of the line).

Two-point Form

$$y - y_1 = \frac{y_2 - y_1}{x_2 - x_1}(x - x_1)$$

where (x_1, y_1) and (x_2, y_2) are two points on the line with $x_2 \neq x_1$. This is equivalent to the point-slope form, where the slope is explicitly given as $(y_2 - y_1)/ (x_2 - x_1)$.

Intercept Form

$$\frac{x}{a} + \frac{y}{b} = 1$$

where a and b must be nonzero. The graph of the equation has x-intercept a and y-intercept b. The intercept form can be converted to the standard form by setting A = 1/a, B = 1/b and C = 1.

Parametric Form

$$x = Tt + U$$

and

$$y = Vt + W$$

Two simultaneous equations in terms of a variable parametre t, with slope m = V/ T, x-intercept (VU–WT)/ V and y-intercept (WT–VU)/ T.

This can also be related to the two-point form, where,

T = p–h, U = h, V = q–k, and W = k:

$$x = (p-h)t + h$$

and

$$y = (q-k)t + k$$

In this case t varies from 0 at point (h,k) to 1 at point (p,q), with values of t between 0 and 1 providing interpolation and other values of t providing extrapolation.

Polar Form

$$r = \frac{mr\cos\theta + b}{\sin}$$

where m is the slope of the line and b is the y-intercept. When θ = 0 the graph will be undefined. Thus, the equation can be rewritten to eliminate discontinuities:

$$r \sin \theta = mr \cos \theta + b$$

Normal Form

$$y\sin\phi + x\cos\phi - p = 0$$

where ϕ is the angle of inclination of the normal and p is the length of the normal. The normal is defined to be the shortest segment between the line in question and the origin. Normal form can be derived from general form by dividing all of the coefficients by,

$$\frac{|C|}{-C}\sqrt{A^2 + B^2}$$

This form is also called the Hesse standard form, after the German mathematician Ludwig Otto Hesse.

Special Cases

$$y = b$$

This is a special case of the standard form where A = 0 and B = 1, or of the slope-intercept form where the slope M = 0. The graph is a horizontal line with y-intercept equal to b. There is no x-intercept, unless b = 0, in which case the graph of the line is the x-axis, and so every real number is an x-intercept.

$$x = a$$

This is a special case of the standard form where A = 1

and $B = 0$. The graph is a vertical line with x-intercept equal to a. The slope is undefined. There is no y-intercept, unless $a = 0$, in which case the graph of the line is the y-axis, and so every real number is a y-intercept.

$$y = y \text{ and } x = x$$

In this case all variables and constants have canceled out, leaving a trivially true statement. The original equation, therefore, would be called an identity and one would not normally consider its graph (it would be the entire xy-plane). An example is $2x + 4y = 2(x + 2y)$. The two expressions on either side of the equal sign are always equal, no matter what values are used for x and y.

$$e = f$$

In situations where algebraic manipulation leads to a statement such as $1 = 0$, then the original equation is called inconsistent, meaning it is untrue for any values of x and y (i.e. its graph would be the empty set) An example would be $3x + 2 = 3x - 5$.

Connection with Linear Functions and Operators

In all of the named forms (assuming the graph is not a vertical line), the variable y is a function of x, and the graph of this function is the graph of the equation. In the particular case that the line crosses through the origin, if the linear equation is written in the form $y = f(x)$ then f has the properties:

$$f(x+y) = f(x) + f(y)$$

and

$$f(ax) = af(x)$$

where a is any scalar. A function which satisfies these properties is called a linear function, or more generally a linear map. This property makes linear equations particularly easy to solve and reason about.

Note that not all linear equations are linear functions. A linear function preserves the rules of scalars. In other words, equations that have a zero y-intercept are also linear functions, but equations that have non-zero y-intercepts are not due to the fact that there is no possible way to write a matrix scalar to form the accepted slope form.

LINEAR EQUATIONS IN MORE THAN TWO VARIABLES

A linear equation can involve more than two variables. The general linear equation in n variables is:

$$a_1x_1 + a_2x_2 + ... + a_nx_n = b$$

In this form, a_1, a_2, ..., an are the coefficients, x_1, x_2, ..., xn are the variables, and b is the constant. When dealing with three or fewer variables, it is common to replace x_1 with just x, x_2 with y, and x_3 with z, as appropriate.

Such an equation will represent an (n–1)-dimensional hyperplane in n-dimensional Euclidean space.

QUADRATIC EQUATION

A quadratic equation is an equation where the highest power of x is x^2. There are various methods of solving quadratic equations.

Note: If $x^2 = 36$, then x = +6 or –6 (since squaring either of these numbers will give 36). However, 36 = + 6 only.

COMPLETING THE SQUARE

9 and 25 can be written as 3^2 and 5^2 whereas 7 and 11 cannot be written as the square of another exact number. 9 and 25 are called perfect squares. Another example is (9/4) = $(3/2)^2$. In a similar way, $x^2 + 2x + 1 = (x + 1)^2$.

To make $x^2 + 6x$ into a perfect square, we add $(6^2/4) = 9$. The resulting expression, $x^2 + 6x + 9 = (x + 3)^2$ and so is a perfect square. This is known as completing the square. To complete the square in this way, we take the number before the x, square it, and divide it by 4. This technique can be used to solve quadratic equations, as demonstrated in the following example.

Example:

Solve $x^2 - 6x + 2 = 0$ by completing the square,

$$x^2 - 6x = -2$$

To complete the square on the LHS (left hand side), we must add $6^2/4 = 9$. We must, of course, do this to the RHS also.

$$x^2 - 6x + 9 = 7$$

$$(x-3)^2 = 7$$

Now take the square root of each side,

$$x - 3 = \pm\ 2.646 \text{ (the square root of 7 is +2.646 or –2.646)}$$

$$x = 5.646 \text{ or } 0.354$$

Completing the square can also be used to find the maximum or minimum point on a graph.

Example: Find the minimum of the graph $y = 3x^2 - 6x - 3$.

In this case, the x^2 has a'3' in front of it, so we start by taking the three out: $y = 3(x^2 - 2x - 1)$. [This is the same since multiplying it out gives $3x^2 - 6x - 3$]

Now complete the square for the bit in the bracket:

$$y = 3[(x-1)^2 - 2]$$

Multiply out the big bracket:

$$y = 3(x-1)^2 - 6$$

We are trying to find the minimum value that this graph can be. $(x - 1)^2$ must be zero or positive, since squaring a number always gives a positive answer. So the minimum value will occur when $(x - 1)^2 = 0$, which is when $x = 1$. When $x = 1$, $y = -6$. So the minimum point is at (1, –6).

Some people don't like the method of completing the square to solve equations and an alternative is to use the quadratic formula. This is actually derived by completing the square.

The Quadratic Formula

$$x = \frac{-b \pm \sqrt{b^2 - 4ac}}{2a}$$

Where the equation is $ax^2 + bx + c = 0$

Example:

Solve $3x^2 + 5x - 8 = 0$

$$x = \frac{-5 \pm (5^2 - 4\times3\times(-8))}{6}$$

$$= \frac{-5 \pm (25 + 96)}{6}$$

$$= \frac{-5 \pm (121)}{6}$$

$$= -5 + 11 \text{ or } -5 - 11$$

6 6

$$x = 1 \text{ or } -2.66$$

Factorising: Sometimes, quadratic equations can be solved by factorising. In this case, factorising is probably the easiest way to solve the equation.

Example:

$$\text{Solve } x^2 + 2x - 8 = 0$$

$$(x - 2)(x + 4) = 0$$

$$\text{either } x - 2 = 0 \text{ or } x + 4 = 0$$

$$x = 2 \text{ or } x = -4$$

SIMULTANEOUS EQUATIONS

In mathematics, simultaneous equations are a set of equations containing multiple variables. This set is often referred to as a system of equations. A solution to a system of equations is a particular specification of the values of all variables that simultaneously satisfies all of the equations. To find a solution, the solver needs to use the provided equations to find the exact value of each variable. Generally the solver uses either a graphical method, the matrix method, the substitution method, or the elimination method. The elimination method as the addition method, since it involves adding equations (or constant multiples of the said equations) to one another.

This is a set of linear equations, also known as a linear system of equations:

$$\begin{cases} 2x + y = 8 \\ x + y = 6 \end{cases}$$

Solving this involves subtracting $x + y = 6$ from $2x + y = 8$ (using the elimination method) to remove the y-variable, then simplifying the resulting equation to find the value of x, then substituting the x-value into either equation to find y.

The solution of this system is:

$$\begin{cases} x = 2 \\ y = 4 \end{cases}$$

which can also be written as an ordered pair (2, 4), representing on a graph the coordinates of the point of intersection of the two lines represented by the equations.

FINDING SOLUTIONS

Sometimes not all variables can be solved for, and so an answer for at least one variable must be expressed in terms of other variables and so the set of all solutions is infinite; this is typical for the case where the system has fewer equations than variables.

If the number of equations is the same as the number of variables, then probably the system is exactly solvable in the sense that the set of its solutions is infinite; for a system of linear equations in this case there is exactly one solution, for other systems to have several solutions is also typical. Sometimes a system has no solution; this is typical for the case where the system has more equations than variables. If these rules about connection between number of solutions and numbers of equations and variables do not hold, then such situation is often referred to as dependence between equations or between their left parts.

For instance, this occurs in linear systems if one equation is a simple multiple of the other (representing the same line, e.g. $2x + y = 3$ and $4x + 2y = 6$) or if the ratio of like variables in two linear equations is the same (representing parallel lines, e.g. $2x + y = 3$ and $6x + 3y = 7$ where the ratio of comparable letters is 3).

Systems of two equations in two real-value unknowns usually appear as one of five different types, having a relationship to the number of solutions:

- Systems that represent intersecting sets of points such as lines and curves, and that are not of one of the types. This can be considered the normal type, the others being exceptional in some respect. These systems usually have a finite number of solutions, each formed by the coordinates of one point of intersection.
- Systems that simplify down to false (for example, equations such as $1 = 0$). Such systems have no points of intersection and no solutions. This type is found, for example, when the equations represent parallel lines.

- Systems in which both equations simplify down to an identity (for example, x = 2x ? x and 0y = 0). Any assignment of values to the unknown variables satisfies the equations. Thus, there are an infinite number of solutions: all points of the plane.
- Systems in which the two equations represent the same set of points: they are mathematically equivalent (one equation can typically be transformed into the other through algebraic manipulation). Such systems represent completely overlapping lines, or curves, etc. One of the two equations is redundant and can be discarded. Each point of the set of points corresponds to a solution. Usually, this means there are an infinite number of solutions.
- Systems in which one (and only one) of the two equations simplifies down to an identity. It is therefore redundant, and can be discarded, as per the previous type. Each point of the set of points represented by the other equation is a solution of which there are then usually an infinite number.

The equation x2 + y2 = 0 can be thought of as the equation of a circle whose radius has shrunk to zero, and so it represents a single point: (x = 0, y = 0), unlike a normal circle containing an infinity of points. This and similar examples show the reason why the last two types need the qualification"usually". An example of a system of equations of the first type with an infinite number of solutions is given by $x = |x|$, $y = |y|$ (where the notation $|o|$ denotes the absolute value function), whose solutions form a quadrant of the x-y plane. Another example is $x = |y|$, $y = |x|$, whose solution represents a ray.

SUBSTITUTION METHOD

Systems of simultaneous equations can be hard to solve unless a systematic approach is used. A common technique is the substitution method: Find an equation that can be written with a single variable as the subject, in which the left-hand side variable does not occur in the right-hand side expression. Next, substitute that expression where that variable appears in the other equations, thereby obtaining a smaller system with fewer variables. After that smaller system has been solved (whether by further application of the substitution method or by other methods).

In this set of equations,

$$\begin{cases} x^2 + y^2 = 1 \\ 2x = 4y = 0 \end{cases}$$

x is made the subject of the second equation:

$$x = -2y$$

then, this result is substituted into the first equation:

$$(-2y)^2 + y^2 = 1$$

After simplification, this yields the solutions,

$$y = \pm\sqrt{\frac{1}{5}}$$

and by substituting this in x = –2y the corresponding x values are obtained.

The two solutions of the system of equations are then:

$$x = -2\sqrt{\frac{1}{5}}, y = \sqrt{\frac{1}{5}}$$

and

$$x = 2\sqrt{\frac{1}{5}}, y = -\sqrt{\frac{1}{5}}$$

ELIMINATION METHOD

Elimination by judicious multiplication is the other commonly used method to solve simultaneous linear equations. It uses the general principles that each side of an equation still equals the other when both sides are multiplied (or divided) by the same quantity, or when the same quantity is added (or subtracted) from both sides. As the equations grow simpler through the elimination of some variables, a variable will eventually appear in fully solvable form, and this value can then be"back-substituted" into previously derived equations by plugging this value in for the variable. Typically, each"back-substitution" can then allow another variable in the system to be solved.

MATRICES

Systems of equations may also be represented in terms of matrices, allowing various principles of matrix operations to be handily applied to the problem.

Systems of simultaneous linear equations are studied in linear algebra; they are solved using Gaussian elimination or the Cholesky decomposition.

To determine approximate solutions to general systems numerically on a computer, the n-dimensional Newton's method may be used. Algebraic geometry is essentially the theory of simultaneous polynomial equations.

The question of effective computation with such equations belongs to elimination theory. Simultaneous equation models are a form of statistical model in the form of a set of linear simultaneous equations. They are often used in econometrics.

In modular arithmetic, simple systems of simultaneous congruences can be solved by the method of successive substitution. Simultaneous equations are easier to solve using this method.

LEAST-SQUARES

A set of linear simultaneous equations can be written in matrix form as Ax = y. If there are more equations than variables, the system is called overdetermined, and has (in general) no solutions.

The system can then be changed to $(A^TA)x = A^Ty$. The new system has as many equations as variables (the matrix ATA is a square matrix) and can be solved in the usual way.

The solution is a least-squares solution of the original, overdetermined system, minimizing the Euclidean norm $||Ax - y||$, a measure of the discrepancy between the two sides in the original system.

MATRIX ALGEBRA

WHAT IS A MATRIX?

A matrix is a rectangular array of numbers (called elements), consisting of m rows and n columns:

$$A = \begin{pmatrix} a_{1,1} & a_{1,2} & \cdot & \cdot & a_{1,n} \\ a_{2,1} & a_{2,2} & \cdot & \cdot & a_{2,n} \\ \cdot & \cdot & & & \cdot \\ \cdot & \cdot & & & \cdot \\ a_{m,1} & a_{m,2} & \cdot & \cdot & a_{m,n} \end{pmatrix}$$

This is said to be a matrix of order m by n. For instance, here is a 2 by 3 matrix:

$$\begin{pmatrix} 1 & 2 & 3 \\ 3 & 2 & 1 \end{pmatrix}$$

If you are from a C/C++ programming background, you wiii notice that a matrix is very much like a 2 dimensional array, but beware - the indices start from 1 (whereas C array indices start from zero). Also, note that the ordering is row index followed by column index, where a programmer might naturally put the column index first. This is just the way matrix algebra is traditionally written. Matrices are useful in a solving a number of problems. As suggested, go on to give a couple of real life examples.

MATRIX ALGEBRA

Addition

It is only possible to add 2 matrices if they are of the same order (i.e., same number of rows and columns). If we add matrices A and B to give a result X, then each element of X is simply the sum of the 2 corresponding elements of A and B,

$$x_{i,j} = a_{i,j} + b_{i,j}$$

for example:

$$A = \begin{pmatrix} 1 & 2 & 3 \\ 3 & 2 & 1 \end{pmatrix}$$

$$B = \begin{pmatrix} 0 & -1 & 0 \\ 1 & 0 & 1 \end{pmatrix}$$

$$A + B = \begin{pmatrix} 1 & 1 & 3 \\ 5 & 5 & 7 \end{pmatrix}$$

Since matrix addition is performed simply by adding the individual elements, clearly you will get the same result whatever order you add the matrices in.

$$A + B = B + A$$

$$(A+B) + C = A + (B + C)$$

Subtraction

Subtraction of 2 matrices is analogous with addition:

$$x_{i,j} = a_{i,j} - b_{i,j}$$

for example, using the same A and B as before,

$$A - B = \begin{pmatrix} 1 & 3 & 3 \\ 3 & 5 & 5 \end{pmatrix}$$

Scalar Multiplication

Scalar multiplication, ie multiplying a matrix by á number (e.g. F) is simply a matter of multiplying each element of the matrix by the number:

$$x_{i,j} = F \times a_{j,i}$$

For example:

$$3 \times A = 3 \times \begin{pmatrix} 1 & 2 & 3 \\ 4 & 5 & 6 \end{pmatrix} \begin{pmatrix} 3 & 6 & 9 \\ 12 & 15 & 18 \end{pmatrix}$$

It should be clear from the scalar multiplication is commutative, ie

$$3 \times A = A \times 3$$

and also that scalar multiplication is distributive over addition and subtraction, ie.

$$3 \times (A + B) = A + 3 \times B$$

Matrix Multiplication

We can multiply 2 matrices to give a matrix result:

$$X = A \times B$$

It is only possible to multiply A and B if the number of columns of A is equal to the number of rows of B (A and B are then said to be conformable).

If A is an n by m matrix, and B is an m by p matrix, then X will be a n by p matrix.

The definition of X is:

$$x_{i,j} = \sum_{k=1}^{m} a_{i,k} \times b_{k,j}$$

For example, we can multiply a 2 by 3 matrix and a 3 by 2 matrix, resulting in a 2 by 2 matrix:

$$\begin{pmatrix} 1 & 3 & 5 \\ 7 & 9 & 11 \end{pmatrix} \times \begin{pmatrix} 0 & 6 \\ 2 & 8 \\ 4 & 10 \end{pmatrix} =$$

$$\begin{pmatrix} 1\times0+3\times2+5\times4 & 1\times6+3\times8+5\times10 \\ 7\times0+9\times2+11\times4 & 7\times6+9\times8+11\times10 \end{pmatrix} = \begin{pmatrix} 26 & 80 \\ 62 & 224 \end{pmatrix}$$

Notice what happens if we change the order of the 2 matrices. This time we are multiplying a 3 by 2 matrix with a 2 by 3 matrix, and the result is a 3 by 3 matrix::

$$\begin{pmatrix} 0 & 6 \\ 2 & 8 \\ 4 & 10 \end{pmatrix} \begin{pmatrix} 1 & 3 & 5 \\ 7 & 9 & 11 \end{pmatrix} \times \begin{pmatrix} 42 & 54 & 66 \\ 58 & 78 & 98 \\ 74 & 102 & 130 \end{pmatrix}$$

This emphasizes that matrix multiplication is not commutative. In fact, if you exchange the matrices, the multiplication may become invalid due to conformability. For example, if A is a 2 by 3 matrix and B is a 3 by 3 matrix, it is possible to for the product AB, but not BA.

Matrix multiplication is, however, associative and distributive.

In summary:

$$A \times B \neq B \times A$$

$$(A \times B) \times C = A \times (B \times C)$$

$$(A + B) \times C = A \times C + B \times C$$

Transposition

Transposing a matrix means converting and m by n matrix into an n by m matrix, by"flipping" the rows and columns.

$$x_{i,j} = a_{j,i}$$

It is denoted by a superscript T, eg:

$$A = \begin{pmatrix} 1 & 2 & 3 \\ 4 & 5 & 6 \end{pmatrix}$$

$$A^T = \begin{pmatrix} 1 & 4 \\ 2 & 5 \\ 3 & 6 \end{pmatrix}$$

As an aside, there is an interesting relationship between transposition and multiplication:

$$(A \times B)^T = B^T \times A^T$$

If you are interested, you can prove this for yourself fairly easily. Hint - look at the definition of matrix multiply, and try swapping the subscripts!

Equality

2 matrices are considered to be equal if they are of the same order, and if all their corresponding elements are equal.

BINOMIAL THEOREM

We know that.

$$(x+y)^0 = 1$$

$$(x+y)^1 = x+y$$

$$(x+y)^2 = x^2 + 2xy + y^2$$

and we can easily expand

$$(x+y)^3 = x^3 + 3x^2y + 3xy^2 + y^3$$

For higher powers, the expansion gets very tedious by hand! Fortunately, the Binomial Theorem gives us the expansion for any positive integer power of (x + y):

For any positive intger n,

$$(x+y)^n = \sum_{k=0}^{n} \binom{n}{k} x^{n-k} y^k$$

where

$$\binom{n}{k} = \frac{(n)(n-1)(n-2)\cdots(n-(k-1))}{k!} = \frac{n!}{k!(n-k)!}$$

Example

By the Binomial Theorem,

$$(x+y)^3 = \sum_{k=0}^{3} \binom{3}{k} x^{3-k} y^k$$

$$= \binom{3}{0}x^3 + \binom{3}{1}x^2y + \binom{3}{2}xy^2 + \binom{3}{3}y^3$$

$$= x^3 + 3x^2y + 3xy^2 + y^3$$

as expected.

EXTENSIONS OF THE BINOMIAL THEOREM

A useful special case of the Binomial Theorem is,

$$(1+x)^n \sum_{k=0}^{n} \binom{n}{k} x^k$$

for any positive integer n, which is just the Taylor series for $(1 + x)^n$.

This formula can be extended to all real powers α:

$$(1+x)^\alpha \sum_{k=0}^{\infty} \binom{\alpha}{k} x^k$$

for any real number α, where,

$$\binom{\alpha}{k} = \frac{(\alpha)(\alpha-1)(\alpha-2)\cdots(\alpha-(k-1))}{k!} = \frac{\alpha!}{k!(\alpha-k)!}$$

But for $|x| \ll 1$ higher powers of x get small very quickly, so,

$$(1+x)^\alpha$$

can be approximated to any accuracy we need by truncating the series after a _nite number of terms.

KEY CONCEPTS

Binomial Theorem

For any positive integer n,

$$(x+y)^n \sum_{k=0}^{n} \binom{n}{k} x^{n-k} y^k$$

where,

$$\binom{n}{k} = \frac{n(n-1)(n-2)\cdots(n-k+1)}{k!} = \frac{n!}{k!(n-k)!}.$$

PRINCIPLES OF MATHEMATICAL INDUCTION

MATHEMATICAL INDUCTION

Mathematical induction is a method of mathematical proof typically used to establish that a given statement is true of all natural numbers. It is done by proving that the first statement in the infinite sequence of statements is true, and then proving that if any one statement in the infinite sequence of statements is true, then so is the next one.

The method can be extended to prove statements about more general well-founded structures, such as trees; this generalization, known as structural induction, is used in mathematical logic and computer science. Mathematical induction in this extended sense is closely related to recursion.

Mathematical induction should not be misconstrued as a form of inductive reasoning, which is considered non-rigorous in mathematics. In fact, mathematical induction is a form of rigorous deductive reasoning.

In 370 BC, Plato's Parmenides may have contained an early example of an implicit inductive proof. The earliest implicit traces of mathematical induction can be found in Euclid's proof that the number of primes is infinite and in Bhaskara's"cyclic method". An opposite iterated technique, counting down rather than up, is found in the Sorites paradox, where one argued that if 1,000,000 grains of sand formed a heap, and removing one grain from a heap left it a heap, then a single grain of sand (or even no grains) forms a heap.

The earliest implicit proof by mathematical induction for arithmetic sequences was introduced in the al-Fakhri written by al-Karaji around 1000 AD, who used it to prove the binomial theorem, Pascal's triangle, and the sum formula for integral cubes. The sum formula for integral cubes is the (true) proposition that every integer can be expressed by the sum of cubed natural numbers. It is a particular case of what is referred to as Waring's Problem. His proof was the first to make use of the two basic components of an inductive proof. First, he notes the truth of the statement for n = 1. That is, 1 is the sum of a single cube because 1 = 13. Secondly, he derives the truth for n = k from that of n = k – 1. For example, when n = 2, it is true that 2 = 13 + 13. When n = 3, it is true that 3 = 13 + 13 + 13. The truth of the statement can be extrapolated in this way without limit. Of course, as n grows larger, some of the sums of 13 can be rewritten as the cubes of other natural numbers: for example when n=8 then 8 = 23 = [13 × 8]. Of course, this second component is not explicit since, in some sense, al-Karaji's argument is in reverse; this is, he starts from n = 10 and goes down to 1 rather than proceeding upward."

Shortly afterwards, Ibn al-Haytham (Alhazen) used the inductive method to prove the sum of fourth powers, and by extension, the sum of any integral powers. He only stated it for particular integers, but his proof for those integers was by induction and generalizable.

Ibn Yahy ā al-Maghrib ī al-Samaw'al came closest to a modern proof by mathematical induction in pre-modern times, which he used to extend the proof of the binomial theorem and Pascal's triangle given by al-Karaji. Al-Samaw'al's inductive argument was only a short step from the full inductive proof of the general binomial theorem. Earliest rigorous use of induction and illustrations of proofs using it was made by Gersonides.

None of these ancient mathematicians, however, explicitly stated the inductive hypothesis. Another similar case (contrary to what Vacca has written, as Freudenthal carefully showed) was that of Francesco Maurolico in his Arithmeticorum libri duo (1575), who used the technique to prove that the sum of the first n odd integers is n2. The first explicit formulation of the principle of induction was given by Pascal in his Traité du triangle arithmétique (1665). Another Frenchman, Fermat, made ample use of a related principle, indirect proof by infinite descent. The inductive hypothesis was also employed by the Swiss Jakob Bernoulli, and from then on it became more or less well known. The modern rigorous and systematic treatment of the principle came only in the 19th century, with George Boole, Giuseppe Peano.

DESCRIPTION

The simplest and most common form of mathematical induction proves that a statement involving a natural number n holds for all values of n.

The proof consists of two steps:

- *The basis (base case)*: showing that the statement holds when n is equal to the lowest value that n is given in the question. Usually, n = 0 or n = 1.
- *The inductive step*: showing that if the statement holds for some n, then the statement also holds when n + 1 is substituted for n.

The assumption in the inductive step that the statement holds for some n is called the induction hypothesis (or inductive hypothesis). To perform the inductive step, one assumes the induction hypothesis and then uses this assumption to prove the statement for n + 1.

The choice between n = 0 and n = 1 in the base case is specific to the context of the proof: If 0 is considered a natural number, as is common in the fields of combinatorics and mathematical logic, then n = 0. If, on the other hand, 1 is taken as the first natural number, then the base case is given by n = 1.

This method works by first proving the statement is true for a starting value, and then proving that the process used to go from one value to the next is valid. If these are both proven, then any value can be obtained by performing the process repeatedly. It may be helpful to think of the domino effect; if one is presented with a long row of dominoes standing on end, one can be sure that:

- The first domino will fall
- Whenever a domino falls, its next neighbour will also fall,

so it is concluded that all of the dominoes will fall, and that this fact is inevitable.

Another analogy can be to consider a set of identical lily

pads, all equally spaced in a line across a pond, with the first and last lily pads adjacent to the two sides of the pond.

If a frog wishes to traverse the pond, it must:

- Determine if the first lily pad will hold its weight.
- Prove that it can jump from one lily pad to another.

Thus, it can conclude that it can jump to all of the lily pads, however many lily pads there are, and cross the pond.

AXIOM OF INDUCTION

The basic assumption or axiom of induction is, in logical symbols,

$$[P(0) \wedge (\forall k \in \mathbb{N})(P(k) \Rightarrow P(k+1))] \Rightarrow (\forall n \in \mathbb{N})[P(n)]$$

where P is any proposition and k and n are both natural numbers.

In other words, the basis P(0) being true along with the inductive case ("P(k) is true implies P(k + 1) is true" for all natural k) being true together imply that P(n) is true for any natural number n. A proof by induction is then a proof that these two conditions hold, thus implying the required cease.

This works because k is used to represent an arbitrary natural number. Then, using the inductive hypothesis, i.e. that P(k) is true, show P(k + 1) is also true. This allows us to"carry" the fact that P(0) is true to the fact that P(1) is also true, and carry P(1) to P(2), etc., thus proving P(n) holds for every natural number n.

Note that the first quantifier in the axiom ranges over predicates rather than over individual numbers. This is called a second-order quantifier, which means that the axiom is stated in second-order logic. Axiomatizing arithmetic induction in first-order logic requires an axiom schema containing a separate axiom for each possible predicate. The Peano axioms contains further discussion of this issue.

EXAMPLE

Mathematical induction can be used to prove that the following statement

$$0+1+2+\cdots+n=\frac{n+(n+1)}{2}$$

holds for all natural numbers n. It gives a formula for the sum of the natural numbers less than or equal to number n. The proof that the statement is true for all natural numbers n proceeds as follows.

Call this statement P(n).

Basis: Show that the statement holds for n = 0.

P(0) amounts to the statement:

$$0 = \frac{0 \cdot (0+1)}{2}$$

In the left-hand side of the equation, the only term is 0, and so the left-hand side is simply equal to 0.

In the right-hand side of the equation, $0 \cdot (0 + 1)/2 = 0$.

The two sides are equal, so the statement is true for n = 0. Thus it has been shown that P(0) holds.

Inductive step: Show that if P(n) holds, then also P(n + 1) holds. This can be done as follows.

Assume P(n) holds (for some unspecified value of n). It must then be shown that P(n + 1) holds, that is:

$$(0+1+2+\cdots+n)+(n+1) = \frac{(n+1)((n+1)+1)}{2}$$

Using the induction hypothesis that P(n) holds, the left-hand side can be rewritten to:

$$\frac{n(n+1)}{2}+(n+1) = \frac{(n+1)((n+1)+1)}{2}$$

Algebraically:

$$\frac{n(n+1)}{2}+(n+1) = \frac{(n+1)((n+1)+1)}{2}$$

$$= \frac{(n+1)(n+2)}{2}$$

$$= \frac{(n+1)((n+1)+1)}{2}$$

thereby showing that indeed P(n + 1) holds.

Since both the basis and the inductive step have been proved, it has now been proved by mathematical induction that P(n) holds for all natural n. Q.E.D.

VARIANTS

In practice, proofs by induction are often structured differently, depending on the exact nature of the property to be proved.

Starting at some other Number

If we want to prove a statement not for all natural numbers but only for all numbers greater than or equal to a certain number b then:

- Showing that the statement holds when n = b.
- Showing that if the statement holds for n = m ≥ b then the same statement also holds for n = m + 1.

This can be used, for example, to show that $n^2 \geq 3n$ for $n \geq 3$. A more substantial example is a proof that,

$$\frac{n^n}{3^n} < n! < \frac{n^n}{2^n} \text{ for } m \geq 6$$

In this way we can prove that P(n) holds for all n ≥1, or even n ≥–5. This form of mathematical induction is actually a special case of the previous form because if the statement that we intend to prove is P(n) then proving it with these two rules is equivalent with proving P(n + b) for all natural numbers n with the first two steps.

Building on n = 2

In mathematics, many standard functions, including operations such as"+" and relations such as"=", are binary, meaning that they take two arguments. Often these functions possess properties that implicitly extend them to more than two arguments. For example, once addition a + b is defined and is known to satisfy the associativity property (a + b) + c = a + (b + c), then the trinary addition a + b + c makes sense, either as (a + b) + c or as a + (b + c). Similarly, many axioms and theorems in mathematics are stated only for the binary versions of mathematical operations and relations, and implicitly extend to higher-arity versions.

Suppose that we wish to prove a statement about an n-ary operation implicitly defined from a binary operation, using mathematical induction on n. Then it should come as no surprise that the n = 2 case carries special weight. Here are some examples.

Example: Product Rule for the Derivative

In this example, the binary operation in question is multiplication (of functions). The usual product rule for the derivative taught in calculus states:

$$(fg)' = f'g + g'f$$

or in logarithmic derivative form

$$(fg)' = f'/f + g'/g$$

This can be generalized to a product of n functions. One has,

$$(f_1 f_2 f_3 \cdots f_n)'$$

$$= (f_1 f_2 f_3 \cdots f_n)' + (f_1 f_2' f_3 \cdots f_n) + (f_1 f_2 f_3' \cdots f_n) + \cdots + (f_1 f_2 \cdots f_{n-1} f_n')$$

or in logarithmic derivative form,

$$(f_1 f_2 f_3 \cdots f_n)' / (f_1 f_2 f_3 \cdots f_n)$$

$$(f_1' f_1) + (f_2' / f_2) + (f_3' / f_3) + \cdots + (f_n' / f_n)$$

In each of the n terms of the usual form, just one of the factors is a derivative; the others are not.

When this general fact is proved by mathematical induction, the n = 0 case is trivial, $(1)' = 0$ (since the empty product is 1, and the empty sum is 0). The n = 1 case is also trivial, $f_1' = f_1'$ And for each n ≥ 3, the case is easy to prove from the preceding n ≥ 1 case. The real difficulty lies in the n = 2 case, which is why that is the one stated in the standard product rule.

Example: Pólya's Proof that there is no "Horse of a Different Colour"

In this example, the binary relation in question is an equivalence relation applied to horses, such that two horses are equivalent if they are the same colour. The argument is essentially identical but the crucial n = 1 case fails, causing the entire argument to be invalid.

In the middle of the 20th century, a commonplace colloquial locution to express the idea that something is unexpectedly different from the usual was"That's a horse of a different colour!". George Pólya posed the following exercise: Find the error in the following argument, which purports to prove by mathematical induction that all horses are of the same colour:

- *Basis*: If there is only one horse, there is only one colour.
- *Induction step*: Assume as induction hypothesis that within any set of n horses, there is only one colour. Now look at any set of n + 1 horses. Number them: 1, 2, 3..., n, n + 1. Consider the sets {1, 2, 3..., n} and {2, 3, 4..., n + 1}. Each is a set of only n horses, therefore within each there is only one colour. But the two sets overlap, so there must be only one colour among all n + 1 horses.

Beginning the induction at 0, the n = 0 case is trivial (any horse is the same colour as itself), and the inductive step is correct in all cases n ≥ 2. However, the logic of the inductive step is incorrect when n = 1, because the statement that"the two sets overlap" is false. Indeed, the n = 1 case is clearly the crux of the matter; if one could prove the n = 1 case, then all higher cases would follow from the transitive property of the equivalence relation.

Induction on More than one Counter

It is sometimes desirable to prove a statement involving two natural numbers, n and m, by iterating the induction process. That is, one performs a basis step and an inductive step for n, and in each of those performs a basis step and an inductive step for m. For example, the proof of commutativity accompanying addition of natural numbers. More complicated arguments involving three or more counters are also possible.

Infinite Descent

Another variant of mathematical induction - the method of infinite descent - was one of Pierre de Fermat's Favourites. This method of proof works in reverse, and can assume several slightly different forms. For example, it might begin by showing that if a statement is true for a natural number n it must also be true for some smaller natural number m (m < n). Using mathematical induction (implicitly) with the inductive hypothesis being that the statement is false for all natural numbers less than or equal to m, we can conclude that the statement cannot be true for any natural number n.

COMPLETE INDUCTION

Another generalization, called complete induction (or strong induction or course of values induction), that in the second step we may assume not only that the statement holds for n = m but also that it is true for all n less than or equal to m.

In complete induction it is not necessary to list the base case as a separate assumption. When considering the first case, it is vacuously true that the statement holds for all previous cases; the inductive step of complete induction in this situation corresponds to the base case in ordinary induction. Thus the proof then of the inductive step in complete induction needs to be able to work with an empty antecedent; the first proof is not of this kind (but can be converted).

Complete induction is most useful when several instances of the inductive hypothesis are required for each inductive step. For example, complete induction can be used to show that,

$$F(n)=\frac{(\varphi_+)^n-(\varphi_-)^n}{(\varphi_+)^n+(\varphi_-)}$$

where F(n) is the nth Fibonacci number and

$$\varphi_+=\left(1+\sqrt{5}\right)/2$$

(the golden ratio) and,

$$\varphi_-=\left(1-\sqrt{5}\right)/2$$

are the roots of,

$$x^2-x-1=0\cdot$$

By using the definition $F(m+1)=F(m)+F(m-1)$, the identity can be verified by direct calculation for F (m + 1) if we assume that it already holds for both F(m) and F(m–1). To complete the proof, the identity must be verified in the two base cases n = 0 and n = 1. Another proof by complete induction uses the hypothesis that the statement holds for all smaller n more thoroughly. Consider the statement that"every natural number greater than 1 is a product of prime numbers", and assume that for a given m > 1 it holds for all smaller n > 1. If m is prime then it is certainly a product of primes, and if not, then by definition it is a product: m = n_1 n_2, where neither of the factors is equal to 1; hence neither is equal to m, and so both are smaller than m.

The induction hypothesis now applies to n_1 and n_2, so each one is a product of primes. Then m is a product of products of primes; i.e. a product of primes. Note both that the base case (m equal to 2) was never explicitly considered, and that the hypothesis that all smaller numbers than m are products of primes was used, since the factors of m are a priori unknown.

This generalization, complete induction, can be derived from the ordinary mathematical induction. Suppose P(n) is the statement that we intend to prove by complete induction. Let Q(n) mean P(m) holds for all m such that 0 ≥ m ≥ n. Apply mathematical induction to Q(n). Since Q(0) is just P(0), we have the base case.

Now suppose Q(n) is given and we wish to show Q(n+1). Notice that Q(n) is the same as P(0) and P(1) and... and P(n). The hypothesis of complete induction tells us that this implies P(n+1). If we add P(n+1) to Q(n), we get P(0) and P(1) and... and P(n) and P(n+1), which is just Q(n+1). So using mathematical induction, we get that Q(n) holds for all natural numbers n. But Q(n) implies P(n), so we have the cease of strong induction, namely that P(n) holds for all natural numbers n.

Transfinite Induction

The last two steps can be reformulated as one step:

- Showing that if the statement holds for all n < m then the same statement also holds for n = m.

This is in fact the most general form of mathematical induction and it can be shown that it is not only valid for statements about natural numbers, but for statements about elements of any well-founded set, that is, a set with an irreflexive relation < that contains no infinite descending chains.

This form of induction, when applied to ordinals (which form a well-ordered and hence well-founded class), is called transfinite induction. It is an important proof technique in set theory, topology and other fields.

Proofs by transfinite induction typically distinguish three cases:

- When m is a minimal element, i.e. there is no element smaller than m
- When m has a direct predecessor, i.e. the set of elements which are smaller than m has a largest element
- When m has no direct predecessor, i.e. m is a so-called limit-ordinal

Strictly speaking, it is not necessary in transfinite induction to prove the basis, because it is a vacuous special case of the proposition that if P is true of all n < m, then P is true of m. It is vacuously true precisely because there are no values of n < m that could serve as counterexamples.

PROOF OR REFORMULATION OF MATHEMATICAL INDUCTION

The principle of mathematical induction is usually stated as an axiom of the natural numbers. However, it can be proved in some logical systems.

For instance, it can be proved if one assumes:

- The set of natural numbers is well-ordered.
- Every natural number is either zero, or n+1 for some natural number n.
- For any natural number n, n+1 is greater than n.

To derive simple induction from these axioms, we must show that if P(n) is some proposition predicated of n, and if:

- P(0) holds and
- Whenever P(k) is true then P(k+1) is also true then P(n) holds for all n.

We first show that if P(k) is true for all k < m, then P(m) is also true. If m is zero, then P(m) is true. If m = k + 1, then P(k) is true because $k < m$ and so $P(k+1)$ is true which means that $P(m)$ is true. The rest follows from applying the principle of transfinite induction.

OR

Let N = {1, 2, 3, 4...} be the set of natural numbers and P(n) be a mathematical statement involving the natural number n belonging to N such that:

- k is least such that, P(k) is true, i.e., P(n) is true for n = k and not true for any $n < k$.
- P(n + 1) is true whenever P(n) is true, i.e., P(n) is true implies that P(n + 1) is true.

Then P(n) is true for all natural numbers $n > k-1$. Proof: Suppose P(n) is not true for all n in N and $n > k-1$. Then there exists a"r" belongs to N and $r > k-1$ such that P (r) is false. Define S:= {m belongs to N/ P(m) is false}. Since r belongs to S, $S \neq ø$. Hence S is well defined. By axiom of choice, there exists a least element in S.Therefore t belongs to N is the least element in S such that P(t) is false. By our hypothesis if P(t-1) is true then P(t) is true. By contrapositive, we are having P(t-1) is false, as P(t) is false. Hence (t-1) belongs to S. This is the contradiction to t is least element in S. Hence our assumption is wrong. Therefore P(n) is true for any n belongs to N and $n > k-1$.

ARITHMETIC PROGRESSION

An arithmetic progression is a sequence of numbers such that the difference of any two successive members of the sequence is a constant. For example, the sequence 3, 5, 7, 9, 11... is an arithmetic progression with common difference 2.

Arithmetic progression property:

$$a_1 + a_n = a_2 + a_{n-1} = ... = a_k + a_{n-k+1}$$

Formulae for the n-th term can be defined as:

$$a_n = 1/2(a_{n-1} + a_{n+1})$$

If the initial term of an arithmetic progression is a1 and the common difference of successive members is d, then the n-th term of the sequence is given by,

$$a_n = a_1 + (n - 1)d, n = 1, 2...$$

The sum S of the first n values of a finite sequence is given by the formula:

$S = 1/2(a_1 + a_n)n$, where a1 is the first term and an the last.

or

$S = 1/2(2a_1 + d(n-1))n$

ARITHMETIC PROGRESSION PROBLEMS

- Is the row 1,11,21,31... arithemtic progression?
  Solution: Yes it is arithmetic progression with first term 1 and common differnece 10.
- Find the sum of the first 10 numbers from this arithmetic progression 1, 11, 21, 31...
  Solution: we can use this formula $S = 1/2(2a_1 + d(n-1))n$
  $S = 1/2(2.1 + 10(10-1))10 = 5(2 + 90) = 5.92 = 460$
- Can u proof that if the numbers 1/(c + b), 1/(c + a), 1/(a + b) are from arithmetic progression then the numbers a_2, b_2, c_2 are also arithmetic progression.

SUM

The sum of the members of a finite arithmetic progression is called an arithmetic series.

Express the arithmetic series in two different ways:

$$S_n = a_1 + (a_1 + d) + (a_1 + 2d) + \cdots + (a_1 + (n-2)d) + (a_1 + (n-1)d)$$

$$S_n = (a_n - (n-1)d) + (a_n - (n-2)d) + \cdots + (a_n - 2d) + (a_n - d) + a_n$$

Adding both sides of the two equations, all terms involving d cancel:

$$2S_n = n(a_1 + a_n)$$

Rearranging and remembering that an = a_1 + (n – 1)d:

$$S_n = \frac{n}{2}(a_1 + a_n) = \frac{n}{2}[2a_1 + (n-1)d]$$

So, for example, the sum of the terms of the arithmetic progression given by an = 3 + (n-1)(5) up to the 50th term is,

$$S_{50} = \frac{50}{2}[2(3) + (49)(5)] = 6,275.$$

PRODUCT

The product of the members of a finite arithmetic progression with an initial element a1, common differences d, and n elements in total is determined in a closed expression by,

$$a_1 a_2 \cdots a_n = d^n \left(\frac{a_1}{d}\right)^{\overline{n}} = d^n \frac{\Gamma(a_1/d+n)}{\Gamma(a_1/d)}$$

where $x^{\overline{n}}$ denotes the rising factorial and Γ denotes the Gamma function. (Note however that the formula is not valid when a1/ d is a negative integer or zero.)

This is a generalization from the fact that the product of the progression 1×2×...×n is given by the factorial n! and that the product,

$$m \times (m+1) \times (m+2) \times \cdots \times (n-2) \times (n-1) \times n$$

for positive integers m and n is given by

$$\frac{n!}{(m-1)!}$$

Taking the example, the product of the terms of the arithmetic progression given by an = 3 + (n-1)(5) up to the 50th term is,

$$P_{50} = 5^{50} \cdot \frac{\Gamma(3/5+50)}{\Gamma(3/5)} \approx 3.784438 \times 10^{98}$$

GEOMETRIC PROGRESSION

In mathematics, a geometric progression, also known as a geometric sequence, is a sequence of numbers where each term after the first is found by multiplying the previous one by a fixed non-zero number called the common ratio. For example, the sequence 2, 6, 18, 54... is a geometric progression with common ratio 3. Similarly 10, 5, 2.5, 1.25... is a geometric sequence with common ratio 1/2. The sum of the terms of a geometric progression is known as a geometric series.

Thus, the general form of a geometric sequence is,

$$a, ar, ar^2, ar^3, ar^4, \ldots$$

and that of a geometric series is

$$a + ar + ar^2 + ar^3 + ar^4 + ...$$

where $r \neq 0$ is the common ratio and a is a scale factor, equal to the sequence's start value.

ELEMENTARY PROPERTIES

The n-th term of a geometric sequence with initial value a and common ratio r is given by,

$$a_n = ar^{n-1}$$

Such a geometric sequence also follows the recursive relation,

$$a_n = ra_n - 1 \text{ for every integer } n \geq 1$$

Generally, to check whether a given sequence is geometric, one simply checks whether successive entries in the sequence all have the same ratio.

The common ratio of a geometric series may be negative, resulting in an alternating sequence, with numbers switching from positive to negative and back. For instance

1, –3, 9, –27, 81, –243, ...

is a geometric sequence with common ratio –3.

The behaviour of a geometric sequence depends on the value of the common ratio.

If the common ratio is:

- Positive, the terms will all be the same sign as the initial term.
- Negative, the terms will alternate between positive and negative.
- Greater than 1, there will be exponential growth towards positive infinity.
- 1, the progression is a constant sequence.
- Between –1 and 1 but not zero, there will be exponential decay towards zero.
- –1, the progression is an alternating sequence
- Less than –1, for the absolute values there is exponential growth towards infinity.

Geometric sequences (with common ratio not equal to –1,1 or 0) show exponential growth or exponential decay, as opposed to the Linear growth (or decline) of an arithmetic progression such as 4, 15, 26, 37, 48, ... (with common difference 11). This result was taken by T.R. Malthus as the mathematical foundation of his Principle of Population. Note that the two kinds of progression are related: exponentiating each term of an arithmetic progression yields a geometric progression, while taking the logarithm of each term in a geometric progression with a positive common ratio yields an arithmetic progression.

GEOMETRIC SERIES

A geometric series is the sum of the numbers in a geometric progression:

$$\sum_{k=0}^{n} ar^k = ar^0 + ar^1 + ar^2 + ar^3 + \cdots + ar^n.$$

We can find a simpler formula for this sum by multiplying both sides of the equation by 1 – r, and we'll see that,

$$(1-r)\sum_{k=0}^{n} ar^k = (1-r)\left(ar^0 + ar^1 + ar^2 + \cdots + ar^n\right)$$

$$= ar^0 + ar^1 + ar^2 ar^3 + \cdots + ar^n - ar^1$$

$$-ar^2 - ar^3 + \cdots + ar^n$$

$$= a - ar^{n+1}$$

since all the other terms cancel. Rearranging (for $r \neq 1$) gives the convenient formula for a geometric series:

$$\sum_{k=0}^{n} ar^k = \frac{a\left(1-r^{n+1}\right)}{1-r}$$

If one were to begin the sum not from 0, but from a higher term, then,

$$\sum_{k=0}^{n} ar^k = \frac{a\left(r^{n+1}-r^m\right)}{1-r}$$

Differentiating this formula with respect to r allows us to arrive at formulae for sums of the form,

$$\sum_{k=0}^{n} k^s r^k$$

For example:

$$\frac{d}{dr}\sum_{k=0}^{n} r^k = \sum_{k=1}^{n} kr^{k-1} = \frac{1-r^{n+1}}{(1-r)^2} - \frac{(n+1)r^n}{1-r}$$

For a geometric series containing only even powers of r multiply by $1 - r^2$:

$$\left(1-r^2\right)\sum_{k=0}^{n} ar^{2k} = a - ar^{2n+2}$$

Then,

$$\sum_{k=0}^{n} ar^{2k} = \frac{a\left(1-r^{2n+2}\right)}{1-r^2}$$

For a series with only odd powers of r,

$$\left(1-r^2\right)\sum_{k=0}^{n} ar^{2k+1} = ar - ar^{2n+3}$$

and

$$\sum_{k=0}^{n} ar^{2k+1} = \frac{ar - ar^{2n+2}}{1-r^2}$$

Infinite Geometric Series

An infinite geometric series is an infinite series whose successive terms have a common ratio. Such a series converges if and only if the absolute value of the common ratio is less than one ($| r | < 1$). Its value can then be computed from the finite sum formulae,

$$\sum_{k=0}^{\infty} ar^k = \lim_{n\to\infty}\sum_{k=0}^{\infty} ar^k = \lim_{n\to\infty}\frac{a\left(1-r^{n+1}\right)}{1-r} =$$

$$\lim_{n\to\infty}\frac{a}{1-r} - \lim_{n\to\infty}\frac{ar^{n+1}}{1-r}a$$

Since:

$$r+1 \to 0 \text{ as } n \to \infty \text{ when } |r| < 1.$$

Then:

$$\sum_{k=0}^{\infty} ar^k = \frac{a}{1-r} - 0 = \frac{a}{1-r}$$

For a series containing only even powers of r,

$$\sum_{k=0}^{\infty} ar^{2k} = \frac{a}{1-r^2}$$

and for odd powers only,

$$\sum_{k=0}^{\infty} ar^{2k+1} = \frac{ar}{1-r^2}$$

In cases where the sum does not start at k = 0,

$$\sum_{k=m}^{\infty} ar^{k} = \frac{ar^m}{1-r}$$

The formulae given are valid only for $|r| < 1$. The latter formula is valid in every Banach algebra, as long as the norm of r is less than one, and also in the field of p-adic numbers if $|r|_p < 1$. As in the case for a finite sum, we can differentiate to calculate formulae for related sums.

For example,

$$\frac{d}{dr}\sum_{k=0}^{\infty} r^k = \sum_{k=0}^{\infty} kr^{k-1} = \frac{1}{(1-r)^2}$$

This formula only works for $|r| < 1$ as well. From this, it follows that, for $|r| < 1$,

$$\sum_{k=0}^{\infty} kr^k = \frac{r}{(1-r)^2}; \sum_{k=0}^{\infty} k^2r^k = \frac{r(1+r)}{(1-r)^3}; \sum_{k=0}^{\infty} k^3r^3 = \frac{r\left(1+4r+r^2\right)}{(1-r)^4}$$

Also, the infinite series 1/2 + 1/4 + 1/8 + 1/16 + · · · is an elementary example of a series that converges absolutely.

It is a geometric series whose first term is 1/2 and whose common ratio is 1/2, so its sum is

$$\frac{1}{2}+\frac{1}{4}+\frac{1}{8}+\frac{1}{16}+\ldots = \frac{1/2}{1-(+1/2)} = 1$$

The inverse of the series is 1/2 – 1/4 + 1/8 – 1/16 + · · · is a simple example of an alternating series that converges absolutely. It is a geometric series whose first term is 1/2 and whose common ratio is –1/2, so its sum is

$$\frac{1}{2}-\frac{1}{4}+\frac{1}{8}-\frac{1}{16}+\ldots = \frac{1/2}{1-(-1/2)} = \frac{1}{3}.$$

Complex Numbers

The summation formula for geometric series remains valid even when the common ratio is a complex number. In this case the condition that the absolute value of r be less than 1 becomes that the modulus of r be less than 1. It is possible to calculate the sums of some non-obvious geometric series. For example, consider the proposition,

$$\sum_{k=0}^{\infty} \frac{\sin(kx)}{r^k} = \frac{r\sin(x)}{1+r^2-2r\cos(x)}$$

The proof of this comes from the fact that,

$$\sin(kx) = \frac{e^{ikx}-e^{-ikx}}{2i},$$

which is a consequence of Euler's formula. Substituting this into the original series gives,

$$\sum_{k=0}^{\infty} \frac{\sin(kx)}{r^k} = \frac{1}{2i}\left[\sum_{k=0}^{\infty}\left(\frac{e^{ix}}{r}\right)^k - \sum_{k=0}^{\infty}\left(\frac{e^{-ix}}{r}\right)^k\right].$$

This is the difference of two geometric series, and so it is a straightforward application of the formula for infinite geometric series that completes the proof.

PRODUCT

The product of a geometric progression is the product of all terms. If all terms are positive, then it can be quickly computed by taking the geometric mean of the progression's first and last term, and raising that mean to the power given by the number of terms. (This is very similar to the formula for the sum of terms of an arithmetic sequence: take the arithmetic mean of the first and last term and multiply with the number of terms.)

$$\prod_{i=0}^{n} ar^i = \left(\sqrt{a_1 \cdot a_{n+1}}\right)^{n+1} \quad \text{(if } a,r>0\text{)}.$$

Proof:

Let the product be represented by P:

$$P = a \cdot ar \cdot ar^2 \cdots ar^{n-1} \cdot ar^n \cdot$$

Now, carrying out the multiplications, we conclude that,

$$P = a^{n+1} r1 + 2 + 3 + \cdots + (n-1) + (n).$$

Applying the sum of arithmetic series, the expression will yield,

$$P = a^{n+1} r^{\frac{n(n+1)}{2}}.$$

$$P = \left(ar^{\frac{n}{2}} \right) n+1.$$

We raise both sides to the second power:

$$P^2 = \left(a^2 r^n\right)^{n+1} = \left(a \cdot ar^n\right)^{n+1}.$$

Consequently,

$$P^2 = \left(a_1 \cdot a_{n+1}\right)^{n+1}$$

and,

$$P = \left(a_1 \cdot a_{n+1}\right)^{\frac{n+1}{2}},$$

which concludes the proof.

GEOMETRIC PROGRESSION PROBLEMS

Problem 1) Is the row 2, 4, 6, 8... geometric progression?

Solution: No it is not. (2, 4, 8 is a geometric progression)

Problem 2) If we have 2, 4, 8... is geometric progression. What will be the 10-th term?

Solution: We can use the formula an = a1. qn-1

$a_{10} = 2.\ 2^{10-1} = 2.\ 512 = 1024$

Problem 3) Find out the scale factor and the command ratio of a geometric progression if

$a_5 - a_1 = 15$

$a_4 - a_2 = 6$

Solution: there are two geometric progression the first one is with scale factor 1 and common ratio = 2

the second decidion is –16, 1/2

3

Data Analysis

QUANTITATIVE DATA ANALYSIS

If you have decided that a large survey is the most appropriate method to use for your research, by now you should have thought about how you're going to analyse your data.

You will have checked that your questionnaire is properly constructed and worded, you will have made sure that there are no variations in the way the forms are administered and you will have checked over and over again that there is no missing or ambiguous information. If you have a well-designed and well-executed survey, you will minimise problems during the analysis.

COMPUTING SOFTWARE

If you have computing software available for you to use you should find this the easiest and quickest way to analyse your data.

The most common package used by social scientists at this present time is SPSS for windows, which has become increasingly user-friendly over the last few years. However, data input can be a long and laborious process, especially for those who are slow on the keyboard, and, if any data is entered incorrectly, it will influence your results.

Large scale surveys conducted by research companies tend to use questionnaires which can be scanned, saving much time and money, but this option might not be open to you. If you are a student, however, spend some time getting to know what equipment is available for your use as you could save yourself a lot of time and energy by adopting this approach. Also, many software packages at the push of a key produce professional graphs, tables and pie charts which can be used in your final report, again saving a lot of time and effort.

Most colleges and universities provide some sort of statistics course and data analysis course. Or the computing department will provide information leaflets and training sessions on data analysis software.

If you have chosen this route, try to get onto one of these courses, especially those which have a'hands-on' approach as you might be able to analyse your data as part of your course work. This will enable you to acquire new skills and complete your research at the same time.

STATISTICAL TECHNIQUES

For those who do not have access to data analysis software, a basic knowledge of statistical techniques is needed to analyse your data. If your goal is to describe what you have found, all you need to do is count your responses and reproduce them. This is called a frequency count or univariate analysis.

However, there is a problem with missing answers in this type of count. For example, someone might be unwilling to let a researcher know their age, or someone else could have accidentally missed out a question.

If there are any missing answers, a separate'no answer' category needs to be included in any frequency count table. In the final report, some researchers overcome this problem by converting frequency counts to percentages which are calculated after excluding missing data. However, percentages can be misleading if the total number of respondents is fewer than 40.

FINDING A CONNECTION

Although frequency counts are a useful starting point in quantitative data analysis, you may find that you need to do more than merely describe your findings.

Often you will need to find out if there is a connection between one variable and a number of other variables. For example, a researcher might want to find out whether there is a connection between watching violent films and aggressive behaviour. This is called bivariate analysis.

In multivariate analysis the researcher is interested in exploring the connections among more than two variables. For example, a researcher might be interested in finding out whether women aged 40-50, in professional occupations, are more likely to try complementary therapies than younger, non-professional women and men from all categories.

MEASURES OF CENTRAL TENDENCY

This tutorial uses histograms to emphasize different measures of central tendency. A histogram is a type of graph in which the x-axis lists categories or values for a data set, and the y-axis shows a count of the number of cases falling into each category.

For example, if there are 59 men and 48 women in your class, you could represent the information with this histogram:

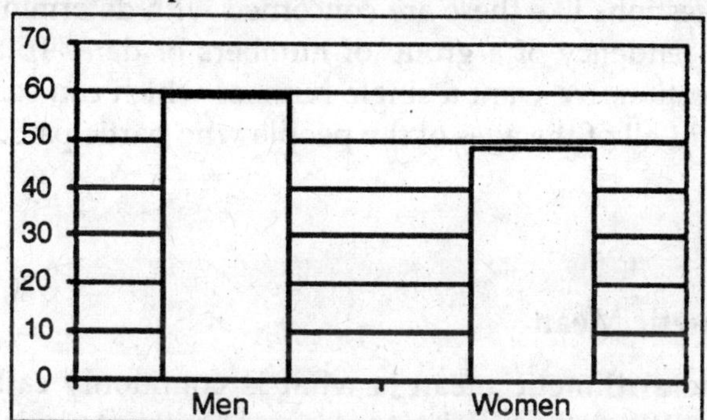

The categories may be non-numeric or may be numeric, as in the following histogram. The x-axis shows the ages for respondents to a survey and the y-axis reports the frequency or count for occurrances of each age.

From the histogram, can you determine what is the"typical" age of the participants in the survey? This question could be answered in several different ways, depending on what you really want to know.

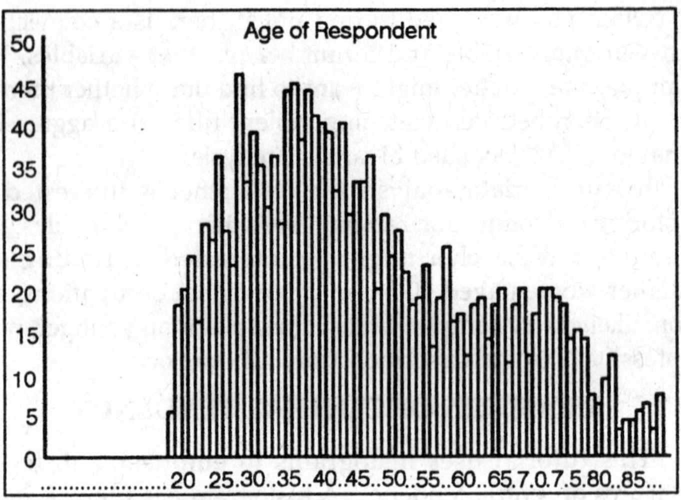

Do you want to determine:

- The average of the ages?
- The age which divides the cases into two equal-sized groups -- the"highs" vs. the"lows"?
- The most common age?

Questions like these are concerned with determining the central tendency of a group of numbers or data. To answer our question, we want a single number which can somehow represent all of the ages of the people who participated in the survey.

MEAN

Arithmetic Mean

The arithmetic mean is what is commonly called the average: When the word"mean" is used without a modifier, it can be assumed that it refers to the arithmetic mean. The mean is the sum of all the scores divided by the number of scores.

The formula in summation notation is:

$$\mu = \Sigma X/N$$

where μ is the population mean and N is the number of scores.

If the scores are from a sample, then the symbol M refers to the mean and N refers to the sample size. The formula for M is the same as the formula for,

μ.

$$M = \Sigma X/N$$

The mean is a good measure of central tendency for roughly symmetric distributions but can be misleading in skewed distributions since it can be greatly influenced by scores in the tail. Therefore, other statistics such as the median may be more informative for distributions such as reaction time or family income that are frequently very skewed

The sum of squared deviations of scores from their mean is lower than their squared deviations from any other number.

For normal distributions, the mean is the most efficient and therefore the least subject to sample fluctuations of all measures of central tendency.

The formal definition of the arithmetic mean is $\mu = E[X]$ where μ is the population mean of the variable X and E[X] is the expected value of X.

The geometric mean is the nth root of the product of the scores. Thus, the geometric mean of the scores: 1, 2, 3, and 10 is the fourth root of 1 × 2 × 3 × 10 which is the fourth root of 60 which equals 2.78.

The formula can be written as:

Geometric mean =

$$\left(\prod X\right)^{\frac{1}{N}}$$

where ΠX means to take the product of all the values of X.

The geometric mean can also be computed by:

- Taking the logarithm of each number
- Computing the arithmetic mean of the logarithms
- Raising the base used to take the logarithms to the arithmetic mean.

X	Ln(X)
1	0
2	0.693147
3	1.098612
10	2.302585
Geometric mean = 2.78	Arithmetic mean = 1.024.
	EXP[1.024] = 2.78

The base of natural logarithms is 2.718. The expression: EXP[1.024] means that 2.718 is raised to the 1.024th power. Ln(X) is the natural log of X.

Naturally, you get the same result using logs base 10 as shown.

X	Log(X)
1	0.0000
2	0.30103
3	0.47712
10	1.00000
Geometric mean = 2.78	Arithmetic mean = 0.44454.
	$10^{0.44454} = 2.78$

If any one of the scores is zero then the geometric mean is zero. The geometric mean does not make sense if any scores are less than zero.

The geometric mean is less affected by extreme values than is the arithmetic mean and is useful as a measure of central tendency for some positively skewed distributions.

The geometric mean is an appropriate measure to use for averaging rates. For example, consider a stock portfolio that began with a value of $1,000 and had annual returns of 13%, 22%, 12%, –5%, and –13%.The table shows the value after each of the five years.

Year	Return	Value
1	13%	1,130
2	22%	1,379
3	12%	1,544
4	-5%	1,467
5	-13%	1,276

The question is how to compute annual rate of return? The answer is to compute the geometric mean of the returns. Instead of using the percents, each return is represented as a multiplier indicating how much higher the value is after the year. This multiplier is 1.13 for a 13% return and 0.95 for a 5% loss. The multipliers for this example are 1.13, 1.22, 1.12, 0.95, and 0.87. The geometric mean of these multipliers is 1.05.

Therefore, the average annual rate of return is 5%. The following table shows how a portfolio gaining 5% a year would end up with the same value ($1,276).

Year	Return	Value
1	5%	1,050
2	5%	1,103
3	5%	1,158
4	5%	1,216
5	5%	1,276

Harmonic Mean

The harmonic mean is used to take the mean of sample sizes. If there are k samples each of size n, then the harmonic mean is defined as:

$$n_h = \frac{k}{\frac{1}{n_1}+\frac{1}{n_2}+\cdots\frac{1}{n_k}}$$

For the numbers 1, 2, 3, and 10, the harmonic mean is:

$$n_h = \frac{k}{\frac{1}{1}+\frac{1}{2}+\frac{1}{3}+\frac{1}{10}} = 2.069$$

= 2.069. This is less than the geometric mean of 2.78 and the arithmetic mean of 4.

MEDIAN

The median is the middle of a distribution: half the scores are the median and half are below the median. The median is less sensitive to extreme scores than the mean and this makes it a better measure than the mean for highly skewed distributions. The median income is usually more informative than the mean income, for example.

The sum of the absolute deviations of each number from the median is lower than is the sum of absolute deviations from any other number.

The mean, median, and mode are equal in symmetric distributions. The mean is typically higher than the median in positively skewed distributions and lower than the median in negatively skewed distributions, although this may not be the case in bimodal distributions.

Computation of Median

When there is an odd number of numbers, the median is simply the middle number. For example, the median of 2, 4, and 7 is 4. When there is an even number of numbers, the median is the mean of the two middle numbers. Thus, the median of the numbers 2, 4, 7, 12 is (4+7)/2 = 5.5.

MODE

The mode is the most frequently occurring score in a distribution and is used as a measure of central tendency. The advantage of the mode as a measure of central tendency is that its meaning is obvious. Further, it is the only measure of central tendency that can be used with nominal data.

The mode is greatly subject to sample fluctuations and is therefore not recommended to be used as the only measure of central tendency. A further disadvantage of the mode is that many distributions have more than one mode. These distributions are called"multi modal." In a normal distribution, the mean, median, and mode are identical.

TRIMEAN

The trimean is computed by adding the 25th percentile plus twice the 50th percentile plus the 75th percentile and dividing by four. What follows is an example of how to compute the trimean. The 25th, 50th, and 75th percentile of the dataset"Example 1" are 51, 55, and 63 respectively.

Therefore, the trimean is computed as:

$$\frac{51+(2)(55)+63}{4}=56$$

The trimean is almost as resistant to extreme scores as the median and is less subject to sampling fluctuations than the arithmetic mean in extremely skewed distributions. It is less efficient than the mean for normal distributions.. The trimean is a good measure of central tendency and is probably not used as much as it should be.

TRIMMED MEAN

A trimmed mean is calculated by discarding a certain percentage of the lowest and the highest scores and then computing the mean of the remaining scores. For example, a mean trimmed 50% is computed by discarding the lower and higher 25% of the scores and taking the mean of the remaining scores.

The median is the mean trimmed 100% and the arithmetic mean is the mean trimmed 0%. A trimmed mean is obviously less susceptible to the effects of extreme scores than is the arithmetic mean.

It is therefore less susceptible to sampling fluctuation than the mean for extremely skewed distributions. It is less efficient than the mean for normal distributions.

Trimmed means are often used in Olympic scoring to minimize the effects of extreme ratings possibly caused by biased judges.

MEASURES OF DISPERSION

While measures of central tendency are used to estimate"normal" values of a dataset, measures of dispersion are important for describing the spread of the data, or its variation around a central value. Two distinct samples may have the same mean or median, but completely different levels of variability, or vice versa. A proper description of a set of data should include both of these characteristics. There are various methods that can be used to measure the dispersion of a dataset, each with its own set of advantages and disadvantages.

RANGE

- Defined as the difference between the largest and smallest sample values.
- One of the simplest measures of variability to calculate.
- Depends only on extreme values and provides no information about how the remaining data is distributed.

STANDARD DEVIATION

- The standard deviation is the square root of the sample variance.
- Defined so that it can be used to make inferences about the population variance.
- Calculated using the formula:

$$SN-1=\sqrt{\frac{1}{N-1}\sum_{i=1}^{N}\left(x_i-\bar{x}\right)^2}$$

- The values computed in the squared term, xi - xbar, are anomalies.
- Not restricted to large sample datsets, compared to the root mean square anomaly.
- Provides significant information into the distribution of data around the mean, approximating normality.
  - The mean ± one standard deviation contains approximately 68% of the measurements in the series.
  - The mean ± two standard deviations contains approximately 95% of the measurements in the series.
  - The mean ± three standard deviations contains approximately 99.7% of the measurements in the series.
- Climatologists often use standard deviations to help classify abnormal climatic conditions. The chart describes the abnormality of a data value by how many standard deviations it is located away from the mean. The probablities in the third column assume the data is normally distributed.

Standard Deviations Away From Mean	**Abnormality**	**Probability of Occurance**
beyond -3 sd	extremely subnormal	0.15%
–3 to –2 sd	greatly subnormal	2.35%
–2 to –1 sd	subnormal	13.5%
–1 to +1 sd	normal	68.0%
+1 to +2 sd	above normal	13.5%
+2 to +3 sd	greatly above normal	2.35%
beyond +3 sd	extremely above normal	0.15%

ROOT MEAN SQUARE ANOMALY/ROOT MEAN SQUARE

Root Mean Square Anomaly

- Also known as root mean square deviation.
- Very similar to standard devation, except used for large sample sizes (i.e., divisior is n instead of n-1) (Devore).
- *RMSA calculated using the formula:*

$$SN = \sqrt{\frac{1}{N}\sum_{i=1}^{N}(x_i - \bar{x})^2},$$

  where x bar is the mean, xi is each data value, and n is the number of observations.
- The term x_i - xbar is an anomaly.
- Provides similar information into the dispersion of data as the standard deviation.
- Often used as a measurement of error.
- More commonly used than the standard deviation function in the statistical analysis of climate data because climate-related datasets are generally quite large in size, in terms of number of data points.

Root Mean Square

- *Calculated using the formula:*

$$x_{rms} = \sqrt{\frac{1}{N}\sum_{i=1}^{N}x_i^2} = \sqrt{\frac{x_1^2 + x_2^2 + \cdots + x_N^2}{N}}$$

- Unlike the RMSA or standard deviation, The mean is not removed in the calculation.
- Acceptable to use only when dealing with large sample datasets (Devore).

INTERQUARTILE RANGE (IQR)

- Calculated by taking the difference between the upper and lower quartiles (the 25th percentile subtracted from the 75th percentile).

- A good indicator of the spread in the centre region of the data.
- Relatively easy to compute.
- More resistant to extreme values than the range.
- Doesn't incorporate all of the data in the sample, compared to the median absolute deviation.
- Also called the fourth-spread.

MEDIAN ABSOLUTE DEVIATION (MAD)

- A more comprehensive alternative to the IQR by incorporating all of the data in the sample.
- MAD = median |Xi - q.5| where Xi represents each value and q.5 represents the median.

TRIMMED VARIANCE

- Similar to variance, except that a proportion of the largest and smallest values in the dataset are ommitted before it is calculated.
- Less affected by outliers since the largest x% and the smallest x% of the sample are eliminated.
- Typical range for x% is 5% to 25%.
- Sometimes multiplied by an adjustment factor to make it more consistant with the ordinary sample variance.
- Analogous to the trimmed mean.

MEAN DEVIATION

The mean deviation is the first measure of dispersion that as suggested, use that actually uses each data value in its computation. It is the mean of the distances between each value and the mean. It gives us an idea of how spread out from the centre the set of values is.

Here's the formula,

Mean Deviation

$$MD = \frac{\sum |x - \bar{x}|}{n}$$

A student took 5 exams in a class and had scores of 92, 75, 95, 90, and 98. Find the mean deviation for her test scores.

First find the mean,

$$\bar{x} = \frac{\sum x}{n}$$

$$= \frac{92+75+95+90+98}{5}$$

$$= \frac{450}{5}$$

$$= 90$$

Now subtract the mean from each score, take the absolute value of each difference, total the absolute values, then divide by the number of values. A column approach works well here.

x	$x - \bar{x}$	$\lvert x - x \rvert$
92	2	2
75	-15	15
95	5	5
90	0	0
98	8	8
$\bar{x} = 90$		$\sum \lvert x - \bar{x} \rvert$ =30

So,

$$MD = \frac{\sum \lvert x - \bar{x} \rvert}{n} = 30 = 6\,.$$

That on the average, this student's test scores deviated by 6 points from the mean.

STANDARD DEVIATION

Standard deviation is a more difficult concept than the others we've covered. And unless you are writing for a specialized, professional audience, you'll probably never use the words"standard deviation" in a story. But that doesn't mean you should ignore this concept.

The standard deviation is kind of the"mean of the mean," and often can help you find the story behind the data. To understand this concept, it can help to learn about what statisticians call normal distribution of data.

A normal distribution of data means that most of the

examples in a set of data are close to the"average," while relatively few examples tend to one extreme or the other.

Let's say you are writing a story about nutrition. You need to look at people's typical daily calorie consumption. Like most data, the numbers for people's typical consumption probably will turn out to be normally distributed. That is, for most people, their consumption will be close to the mean, while fewer people eat a lot more or a lot less than the mean.

When you think about it, that's just common sense. Not that many people are getting by on a single serving of kelp and rice. Or on eight meals of steak and milkshakes. Most people lie somewhere in between.

If you looked at normally distributed data on a graph, it would look something like this:

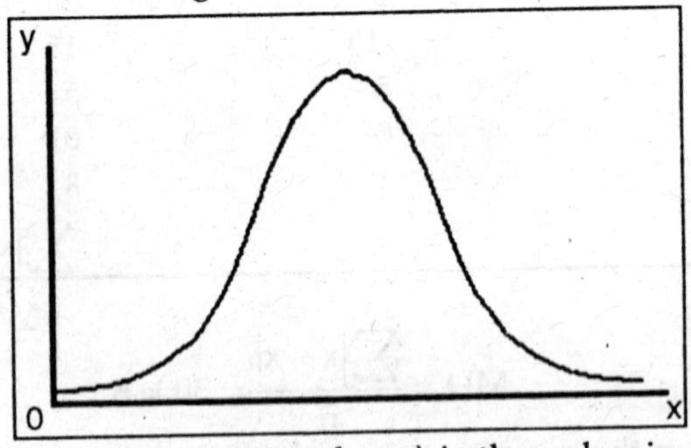

The x-axis (the horizontal one) is the value in question... calories consumed, dollars earned or crimes committed, for example. And the y-axis (the vertical one) is the number of datapoints for each value on the x-axis... in other words, the number of people who eat x calories, the number of households that earn x dollars, or the number of cities with x crimes committed.

Now, not all sets of data will have graphs that look this perfect. Some will have relatively flat curves, others will be pretty steep. Sometimes the mean will lean a little bit to one side or the other. But all normally distributed data will have something like this same"bell curve" shape.

The standard deviation is a statistic that tells you how tightly all the various examples are clustered around the mean in a set of data. When the examples are pretty tightly bunched together and the bell-shaped curve is steep, the standard deviation is small. When the examples are spread apart and the bell curve is relatively flat, that tells you you have a relatively large standard deviation.

Computing the value of a standard deviation is complicated. But let me show you graphically what a standard deviation represents...

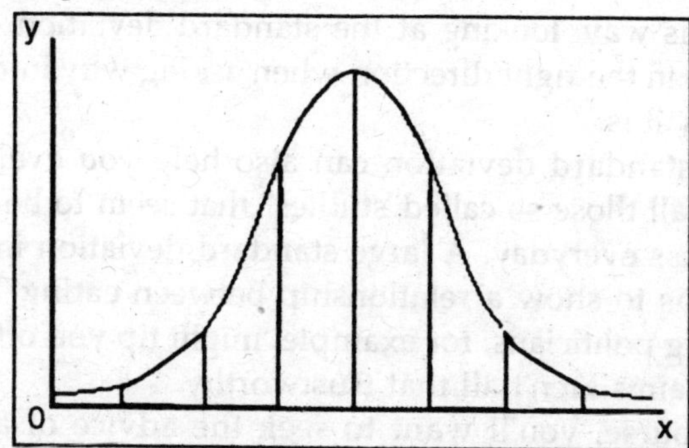

One standard deviation away from the mean in either direction on the horizontal axis accounts for somewhere around 68 per cent of the people in this group. Two standard deviations away from the mean (the red and green areas) account for roughly 95 per cent of the people. And three standard deviations (the red, green and blue areas) account for about 99 per cent of the people.

If this curve were flatter and more spread out, the standard deviation would have to be larger in order to account for those 68 per cent or so of the people. So that's why the standard deviation can tell you how spread out the examples in a set are from the mean.

Why is this useful? Here's an example: If you are comparing test scores for different schools, the standard deviation will tell you how diverse the test scores are for each school.

Let's say Springfield Elementary has a higher mean test score than Shelbyville Elementary. Your first reaction might be to say that the kids at Springfield are smarter.

But a bigger standard deviation for one school tells you that there are relatively more kids at that school scoring towards one extreme or the other. By asking a few follow-up questions you might find that, Springfield's mean was skewed up because the school district sends all of the gifted education kids to Springfield. Or that Shelbyville's scores were dragged down because students who recently have been" main-streamed" from special education classes have all been sent to Shelbyville.

In this way, looking at the standard deviation can help point you in the right direction when asking why information is the way it is.

The standard deviation can also help you evaluate the worth of all those so-called"studies" that seem to be released to the press everyday. A large standard deviation in a study that claims to show a relationship between eating Twinkies and killing politicians, for example, might tip you off that the study's claims aren't all that trustworthy.

Of course, you'll want to seek the advice of a trained statistician whenever you try to evaluate the worth of any scientific research. But if you know at least a little about standard deviation going in, that will make your interview much more productive.

Okay, because so many of you asked nicely...

Here is one formula for computing the standard deviation. A warning, this is for math geeks only! Remember, a decent calculator and stats programme will calculate this for you...

- Terms you'll need to know x = one value in your set of data avg (x) = the mean (average) of all values x in your set of data n = the number of values x in your set of data

For each value x, subtract the overall avg (x) from x, then multiply that result by itself (otherwise known as determining the square of that value). Sum up all those squared values. Then divide that result by (n-1). Got it? Then, there's one more step... find the square root of that last number. That's the standard deviation of your set of data.

Now, remember how one can told you this was one way of computing this? Sometimes, you divide by (n) instead of (n-1). It's too complex to explain here. So don't try to go figuring out a standard deviation. Just be satisified that you've now got a grasp on the basic concept.

BASIC EXAMPLE

Consider a population consisting of the following eight values:

2, 4, 4, 4, 5, 5, 7, 9.

The eight data points have a mean (or average) value of 5:

$$\frac{2+4+4+4+5+5+7+9}{8}=5$$

To calculate the population standard deviation, first compute the difference of each data point from the mean, and square the result:

$$(2-5)^2=(-3)^2=9(5-5)^2=0^2=0$$

$$(4-5)^2=(-1)^2=1(5-5)^2=0^2=0$$

$$(4-5)^2=(-1)^2=1(7-5)^2=2^2=4,$$

$$(4-5)^2=(-1)^2=1(9-5)^2=4^2=16$$

Next divide the sum of these values by the number of values and take the square root to give the standard deviation:

$$\sqrt{\frac{9+1+1+1+0+0+4+16}{8}}=2$$

Therefore, a population standard deviation of 2.

The assumes a complete population. If the 8 values are obtained by random sampling from some parent population, then computing the sample standard deviation would use a denominator of 7 instead of 8.

DEFINITION

Probability Distribution or Random Variable

Let X be a random variable with mean value μ:

$$E[X]=\mu$$

Here the operator E denotes the average or expected value of X. Then the standard deviation of X is the quantity,

$$\sigma=\sqrt{E\left[(X-\mu)^2\right]}.$$

That is, the standard deviation σ (sigma) is the square root of the average value of $(X - \mu)^2$. In the case where X takes random values from a finite data set x1, x_2, x_3...,x_N, with each value having the same probability, the standard deviation is,

$$\sigma = \sqrt{\frac{(x_1 - \mu)^2 + (x_2 - \mu)^2 + \cdots + (x_N - \mu)^2}{N}}.$$

or, using summation notation,

$$\sigma = \sqrt{\frac{1}{N}\sum_{i=1}^{N}(x_i - \mu)^2}.$$

The standard deviation of a (univariate) probability distribution is the same as that of a random variable having that distribution. Not all random variables have a standard deviation, since these expected values need not exist. For example, the standard deviation of a random variable which follows a Cauchy distribution is undefined because its expected value is undefined.

Continuous Random Variable

The standard deviation of a continuous real-valued random variable X with probability density function p(x) is,

$$\sigma = \sqrt{\int (x - \mu)^2 \, p(x)dx},$$

where,

$$\mu = \int x p(x)dx,$$

and where the integrals are definite integrals taken for x ranging over the sample space of X.

In the case of a parametric family of distributions, the standard deviation can be expressed in terms of the parametres. For example, in the case of the log-normal distribution with parametres μ and σ^2, the standard deviation is $[(\exp(\sigma^2)-1)\exp(2\mu+\sigma^2)]1/2$.

ESTIMATION

One can find the standard deviation of an entire population in cases (such as standardized testing) where every member of a population is sampled. In cases where that cannot be done, the standard deviation σ is estimated by examining a random sample taken from the population.

With Standard Deviation of the Sample

An estimator for σ sometimes used is the standard deviation of the sample, denoted by sn and defined as follows:

$$s_n = \sqrt{\frac{1}{N}\sum_{i=1}^{N}(x_i - \bar{x})^2}$$

This estimator has a uniformly smaller mean squared error than the"sample standard deviation" and is the maximum-likelihood estimate when the population is normally distributed. But this estimator, when applied to a small or moderately sized sample, tends to be too low: it is a biased estimator.

The standard deviation of the sample is the same as the population standard deviation of a discrete random variable that can assume precisely the values from the data set, where the probability for each value is proportional to its multiplicity in the data set.

With Sample Standard Deviation

The most common estimator for ? used is an adjusted version, the sample standard deviation, denoted by"s" and defined as follows:

$$s = \sqrt{\frac{1}{N-1}\sum_{i=1}^{N}(x_i - \bar{x})^2}$$

where $\{x_1, x_2, ..., x_N\}$ are the observed values of the sample items and $\bar{x}$ is the mean value of these observations. This correction (the use of N – 1 instead of N) is known as Bessel's correction.

The reason for this correction is that s^2 is an unbiased estimator for the variance s^2 of the underlying population, if that variance exists and the sample values are drawn independently with replacement. However, s is not an unbiased estimator for the standard deviation s; it tends to underestimate the population standard deviation.

Note that the term"standard deviation of the sample" is used for the uncorrected estimator (using N) whilst the term"sample standard deviation" is used for the corrected estimator (using N – 1). The denominator N – 1 is the number of degrees of freedom in the vector of residuals $\left(x_1, \overline{x}, \ldots, x_{N-\overline{x}}\right)$.

Other Estimators

Although an unbiased estimator for σ is known when the random variable is normally distributed, the formula is complicated and amounts to a minor correction. Moreover, unbiasedness, (in this sense of the word), is not always desirable.

IDENTITIES AND MATHEMATICAL PROPERTIES

The standard deviation is invariant to changes in location, and scales directly with the scale of the random variable. Thus, for a constant c and random variables X and Y:

$$\text{stdev}(X+c) = \text{stdev}(X).$$

$$\text{stdev}(cX) = |c|\,\text{stdev}(X).$$

The standard deviation of the sum of two random variables can be related to their individual standard deviations and the covariance between them:

$$\text{stdev}(X+Y) = \sqrt{\text{var}(X) + \text{var}(Y) + 2\,\text{cov}(X,Y)}.$$

where var and cov stand for variance and covariance, respectively. The calculation of the sum of squared deviations can be related to moments calculated directly from the data. In general, we have

$$\text{stdev}(X) = \sqrt{E(X-EX)^2} = \sqrt{E\left(X^2\right) - (EX)^2}$$

For a finite population with equal probabilities on all points, we have

$$\sqrt{\frac{1}{N}\sum_{i=1}^{N}\left(x_i - \overline{x}\right)^2} = \sqrt{\frac{1}{N}\left(\sum_{i-1}^{N} x_i^2\right) - \overline{x}^2}$$

Thus, the standard deviation is equal to the square root of (the average of the squares less the square of the average).

INTERPRETATION AND APPLICATION

A large standard deviation indicates that the data points are far from the mean and a small standard deviation indicates that they are clustered closely around the mean.

For example, each of the three populations {0, 0, 14, 14}, {0, 6, 8, 14} and {6, 6, 8, 8} has a mean of 7. Their standard deviations are 7, 5, and 1, respectively.

The third population has a much smaller standard deviation than the other two because its values are all close to 7. In a loose sense, the standard deviation tells us how far from the mean the data points tend to be.

It will have the same units as the data points themselves. If, for instance, the data set {0, 6, 8, 14} represents the ages of a population of four siblings in years, the standard deviation is 5 years.

As another example, the population {1000, 1006, 1008, 1014} may represent the distances traveled by four athletes, measured in metres. It has a mean of 1007 metres, and a standard deviation of 5 metres.

Standard deviation may serve as a measure of uncertainty. In physical science, for example, the reported standard deviation of a group of repeated measurements should give the precision of those measurements.

When deciding whether measurements agree with a theoretical prediction the standard deviation of those measurements is of crucial importance: if the mean of the measurements is too far away from the prediction (with the distance measured in standard deviations), then the theory being tested probably needs to be revised.

This makes sense since they fall outside the range of values that could reasonably be expected to occur if the prediction were correct and the standard deviation appropriately quantified.

Application Examples

The practical value of understanding the standard deviation of a set of values is in appreciating how much variation there is from the"average" (mean).

Weather

As a simple example, consider average temperatures for cities. While two cities may each have an average temperature of 15 °C, it's helpful to understand that the range for cities near the coast is smaller than for cities inland, which clarifies that, while the average is similar, the chance for variation is greater inland than near the coast.

So, an average of 15 occurs for one city with highs of 25 °C and lows of 5 °C, and also occurs for another city with highs of 18 and lows of 12. The standard deviation allows us to recognize that the average for the city with the wider variation, and thus a higher standard deviation, will not offer as reliable a prediction of temperature as the city with the smaller variation and lower standard deviation.

Sports

Another way of seeing it is to consider sports teams. In any set of categories, there will be teams that rate highly at some things and poorly at others.

Chances are, the teams that lead in the standings will not show such disparity, but will perform well in most categories. The lower the standard deviation of their ratings in each category, the more balanced and consistent they will tend to be. Whereas, teams with a higher standard deviation will be more unpredictable. For example, a team that is consistently bad in most categories will have a low standard deviation. A team that is consistently good in most categories will also have a low standard deviation.

However, a team with a high standard deviation might be the type of team that scores a lot (strong offense) but also concedes a lot (weak defence), or, vice versa, that might have a poor offense but compensates by being difficult to score on. Trying to predict which teams, on any given day, will win, may include looking at the standard deviations of the various team"stats" ratings, in which anomalies can match strengths vs. weaknesses to attempt to understand what factors may prevail as stronger indicators of eventual scoring outcomes. In racing, a driver is timed on successive laps. A driver with a low standard deviation of lap times is more consistent than a driver with a higher standard deviation. This information can be used to help understand where opportunities might be found to reduce lap times.

Finance

In finance, standard deviation is a representation of the risk associated with a given security (stocks, bonds, property, etc.), or the risk of a portfolio of securities (actively managed mutual funds, index mutual funds, or ETFs). Risk is an important factor in determining how to efficiently manage a portfolio of investments because it determines the variation in returns on the asset and/or portfolio and gives investors a mathematical basis for investment decisions (known as mean-variance optimization). The overall concept of risk is that as it increases, the expected return on the asset will increase as a result of the risk premium earned - in other words, investors should expect a higher return on an investment when said investment carries a higher level of risk, or uncertainty of that return. When evaluating investments, investors should estimate both the expected return and the uncertainty of future returns. Standard deviation provides a quantified estimate of the uncertainty of future returns.

For example, let's assume an investor had to choose between two stocks. Stock A over the last 20 years had an average return of 10%, with a standard deviation of 20 percentage points (pp) and Stock B, over the same period, had average returns of 12%, but a higher standard deviation of 30 pp. On the basis of risk and return, an investor may decide that Stock A is the safer choice, because Stock B's additional 2% points of return is not worth the additional 10 pp standard deviation (greater risk or uncertainty of the expected return). Stock B is likely to fall short of the initial investment (but also to exceed the initial investment) more often than Stock A under the same circumstances, and is estimated to return only 2% more on average.

In this example, Stock A is expected to earn about 10%, plus or minus 20 pp (a range of 30% to -10%), about two-thirds of the future year returns. When considering more extreme possible returns or outcomes in future, an investor should expect results of up to 10% plus or minus 60 pp, or a range from 70% to (–)50%, which includes outcomes for three standard deviations from the average return (about 99.7% of probable returns).

Calculating the average return (or arithmetic mean) of a security over a given period will generate an expected return on the asset. For each period, subtracting the expected return from the actual return results in the variance. Square the variance in each period to find the effect of the result on the overall risk of the asset.

The larger the variance in a period, the greater risk the security carries. Taking the average of the squared variances results in the measurement of overall units of risk associated with the asset. Finding the square root of this variance will result in the standard deviation of the investment tool in question.

Population standard deviation is used to set the width of Bollinger bands, a widely adopted technical analysis tool. For example, the upper Bollinger band is given as:

$$\overline{x} + n\sigma_x.$$

Geometric Interpretation

To gain some geometric insights, as suggested, start with a population of three values, x_1, x_2, x_3. This defines a point P = (x_1, x_2, x_3) in R_3. Consider the line L = {(r, r, r): r in R}. This is the"main diagonal" going through the origin. If our three given values were all equal, then the standard deviation would be zero and P would lie on L. So it is not unreasonable to assume that the standard deviation is related to the distance of P to L. And that is indeed the case. To move orthogonally from L to the point P, one begins at the point:

$$M = (\overline{x}, \overline{x}, \overline{x})$$

whose coordinates are the mean of the values we started out with. A little algebra shows that the distance between P and M (which is the same as the orthogonal distance between P and the line L) is equal to the standard deviation of the vector

x_1, x_2, x_3, divided by the square root of the number of dimensions of the vector.

Chebyzshev's Inequality

An observation is rarely more than a few standard deviations away from the mean. Chebyshev's inequality entails the following bounds for all distributions for which the standard deviation is defined.

- At least 50% of the values are within √2 standard deviations from the mean.
- At least 75% of the values are within 2 standard deviations from the mean.
- At least 89% of the values are within 3 standard deviations from the mean.
- At least 94% of the values are within 4 standard deviations from the mean.
- At least 96% of the values are within 5 standard deviations from the mean.
- At least 97% of the values are within 6 standard deviations from the mean.

And in general: At least (1 – 1/k2) × 100% of the values are within k standard deviations from the mean.

Rules for Normally Distributed Data

The central limit theorem says that the distribution of a sum of many independent, identically distributed random variables tends towards the famous"bell-shaped" normal distribution with a probability density function of:

$$\frac{1}{\sqrt{2\pi\sigma^2}}\exp\left(-\frac{(x-\mu)^2}{2\sigma^2}\right)$$

where μ is the arithmetic mean of the sample. The standard deviation therefore is simply a scaling variable that adjusts how broad the curve will be, though also appears in the normalizing constant to keep the distribution normalized for different widths.

If a data distribution is approximately normal then the proportion of data values within z standard deviations of the mean is defined by erf ($z\sigma/\sqrt{2}$). The percentage of data values within z standard deviations of the mean is defined by erf($z\sigma/\sqrt{2}$) × 50% + 50%. If a data distribution is approximately normal then about 68% of the data values are within 1 standard deviation of the mean (mathematically, $\mu \pm \sigma$, where μ is the arithmetic mean), about 95% are within two standard deviations ($\mu \pm 2\sigma$), and about 99.7% lie within 3 standard deviations ($\mu \pm 3\sigma$). This is known as the 68-95-99.7 rule, or the empirical rule.

For various values of z, the percentage of values expected to lie in and outside the symmetric confidence interval CI = ($-z\sigma$, $z\sigma$) are as follows:

$z\sigma$	Percentage within CI	Percentage outside CI	Ratio outside CI
1σ	68.2689492%	31.7310508%	1 / 3.1514871
1.645σ	90%	10%	1 / 10
1.960σ	95%	5%	1 / 20
2σ	95.4499736%	4.5500264%	1 / 21.977894
2.576σ	99%	1%	1 / 100
3σ	99.7300204%	0.2699796%	1 / 370.398
3.2906σ	99.9%	0.1%	1 / 1000
4σ	99.993666%	0.006334%	1 / 15,788
5σ	99.9999426697%	0.0000573303%	1 / 1,744,278
6σ	99.9999998027%	0.0000001973%	1 / 506,800,000
7σ	99.999 999 999 7440%	0.0000000002560%	1 / 390,600,000,000

RELATIONSHIP BETWEEN STANDARD DEVIATION AND MEAN

The mean and the standard deviation of a set of data are usually reported together. In a certain sense, the standard deviation is a"natural" measure of statistical dispersion if the centre of the data is measured about the mean. This is because the standard deviation from the mean is smaller than from any other point. The precise statement is the following: suppose $x_1 ..., x_n$ are real numbers and define the function:

$$\sigma(r)\sqrt{\frac{1}{N-1}\sum_{i=1}^{N}(x_i - r)^2}.$$

Using calculus or by completing the square, it is possible to show that $\sigma(r)$ has a unique minimum at the mean:

$$r = \bar{x}.$$

The coefficient of variation of a sample is the ratio of the standard deviation to the mean. It is a dimensionless number that can be used to compare the amount of variance between populations with means that are close together.

The reason is that if you compare populations with same standard deviations but different means then coefficient of variation will be bigger for the population with the smaller mean. Thus in comparing variability of data, coefficient of variation should be used with care and better replaced with another method.

Often we want some information about the accuracy of the mean we obtained. We can obtain this by determining the standard deviation of the sampled mean. The standard deviation of the mean is related to the standard deviation of the distribution by:

$$\sigma_{mean} = \frac{\sigma}{\sqrt{N}}.$$

where N is the number of observation in the sample used to estimate the mean. This can easily be proven with:

$VAR(X) = \sigma_x^2.$
$VAR(X_1 + X_2) = VAR(X_1) + VAR(X_2),$
$VAR(cX_1) = c^2 VAR(X_1),$
hence,

$$VAR(mean) = VAR\left(\frac{1}{4}\sum_{i=1}^{N} X_i\right) = \frac{1}{N^2} VAR\left(\sum_{i=1}^{N} X_i\right) =$$

$$\frac{1}{N^2}\sum_{i=1}^{N} VAR(X_i) = \frac{N}{N^2} VAR(X) = \frac{1}{N} VAR(X)$$

Resulting in:

$$\sigma_{mean} = \frac{\sigma}{\sqrt{N}}$$

WORKED EXAMPLE

The standard deviation of a discrete random variable is the root-mean-square (RMS) deviation of its values from the mean.

If the random variable X takes on N values $x_1,...,x_N$ (which are real numbers) with equal probability, then its standard deviation σ can be calculated as follows:

- Find the mean, $\overline{x}$, of the values.
- For each value xi calculate its deviation $x_i - \overline{x}$ from the mean.
- Calculate the squares of these deviations.
- Find the mean of the squared deviations. This quantity is the variance σ^2.
- Take the square root of the variance.

This calculation is described by the following formula:

$$\sigma = \sqrt{\frac{1}{N}\sum_{i=1}^{N}(x_i - \overline{x})^2}$$

where $\overline{x}$ *is the arithmetic mean of the values xi, defined as:*

$$\overline{x} = \frac{x_1 + x_2 + \cdots + x_N}{N} = \frac{1}{N}\sum_{i=1}^{N} x_i$$

If not all values have equal probability, but the probability of value xi equals pi, the standard deviation can be computed by:

$$\sigma = \sqrt{\sum_{i=1}^{N} pi(x_i - \overline{x})^2}.$$

where

$$\overline{x} = \sum_{i=1}^{N} p_i x_i.$$

Suppose we wished to find the standard deviation of the distribution placing probabilities 1/4, 1/2, and 1/4 on the points in the sample space 3, 7, and 19.

Step 1: find the probability-weighted mean,

$$3/4 + 7/2 + 19/4 = 9.$$

Step 2: find the deviation of each value in the sample space from the mean,

$$3 - 9 = -6$$
$$7 - 9 = -2$$
$$19 - 9 = 10.$$

Step 3: square each of the deviations, which amplifies large deviations and makes negative values positive,

$$(-6)^2 = 36$$

$$(-2)^2 = 4$$

$$10^2 = 100$$

Step 4: find the probability-weighted mean of the squared deviations,

$$36/4 + 4/2 + 100/4 = 36.$$

Step 5: take the positive square root of the quotient,

$$\sqrt{36} = 6$$

So, the standard deviation of the set is 6. This example also shows that, in general, the standard deviation is different from the mean absolute deviation (which is 5 in this example).

RAPID CALCULATION METHODS

The following two formulas can represent a running (continuous) standard deviation. A set of three power sums $s_{0,1,2}$ are each computed over a set of N values of x, denoted as xk.

$$s_j = \sum_{k=1}^{N} x_k^j$$

Note that s_0 raises x to the zero power, and since x_0 is always 1, s_0 evaluates to N.

Given the results of these three running summations, the values s_0,1,2 can be used at any time to compute the current value of the running standard deviation. This definition for sj can represent the two different phases (summation computation s_j, and σ calculation).

$$\sigma = \frac{1}{s_0}\sqrt{s_0 s_2 - s_1^2}$$

Similarly for sample standard deviation,

$$s = \sqrt{\frac{s_0 s_2 - s_1^2}{s_0 (S_0 - 1)}}$$

In a computer implementation, as the three sj sums become large, we need to consider round-off error, arithmetic overflow, and arithmetic underflow. The method calculates the running sums method with reduced rounding errors:

$$A_0 = 0$$

$$A_i = A_{i-1} + \frac{x_i - A_{i-1}}{i}$$

where A is the mean value.

$$Q_0 = 0$$

$$Q_i = Q_{i-1} + \frac{i-1}{i}(x_i A_{i-1})^2$$

or

$$Q_i = Q_{i-1} + (x_i - A_{i-1})(x_i - A_i)$$

sample variance:

$$s_n^2 = \frac{Q_n}{n-1}$$

standard variance:

$$\sigma_n^2 = \frac{Q_n}{n}.$$

Weighted Calculation

When the values x_i are weighted with unequal weights wi, the power sums s_0,1,2 are each computed as:

$$s_j = \sum_{k=1}^{N} w_k x_k^j$$

And the standard deviation equations remain unchanged. Note that s0 is now the sum of the weights and not the number of samples N. The incremental method with reduced rounding errors can also be applied, with some additional complexity.

A running sum of weights must be computed:

$$W_0 = 0$$

$$W_i = W_{i-1} + w_i$$

and places where 1/i is used must be replaced by wi/Wi:

$$A_0 = 0$$

$$A_i = A_{i-1} + \frac{w_i}{W_i}(x_i - A_{i-1})$$

$$Q_0 = 0$$

$$Q_i = Q_{i-1} + \frac{w_i W_{i-1}}{W}(x_i - A_{i-1})^2 =$$

$$Q_{i-1} + w_i(x_i - A_{i-1})(x_i - A_i)$$

In the final division,

$$\sigma_n^2 = \frac{Q_n}{W_n}$$

and,

$$s_n^2 = \frac{n'}{n'-1}\sigma_n^2$$

where n is the total number of elements, and n' is the number of elements with non-zero weights. The formulas become equal to the simpler formulas given if weights are taken as equal to 1.

COMBINING STANDARD DEVIATIONS

Population-based Statistics

Standard deviations of non-overlapping sub-populations can be aggregated as follows if the size (actual or relative to one another) and means of each are known:

$$\mu X \cup Y = \frac{N_X\mu_X + N_Y\mu_Y}{N_X + N_Y}$$

and

$$\sigma X \cup Y = \sqrt{\frac{N_X\left(\sigma_X^2 + \mu_X^2\right) + N_Y\left(\sigma_Y^2 + \mu_Y^2\right)}{N_X + N_Y} - \mu_{X \cup Y}^2}$$

where

$$X \cap Y \equiv \theta$$

For example, suppose it is known that the average American man has a mean height of 70 inches with a standard deviation of 3 inches and that the average American woman has a mean height of 65 inches with a standard deviation of 2 inches. The mean and standard deviation for American adults could be calculated as:

$$\mu_{\text{height}} \approx \frac{50\% \times 70\text{inches} + 50\% \times 65\text{inches}}{100\%} =$$

$$\frac{70+65}{2}\text{inches} = 67.5\text{inches}$$

$$\sigma_{\text{height}} \approx \sqrt{\frac{\left(3^2+70^2\right)+\left(2^2+65^2\right)}{2} - 67.5^2\text{inches}}$$

$$\sqrt{12.75}\text{inches} \approx 3.5707\text{ inches}$$

For the more general M non-overlapping data sets X1 through XM:

$$\mu\{X1 \cup ... \cup XM\} = \frac{\sum_{i=1}^{M} N_{Xi}\mu X_i}{\sum_{i=1}^{M} N_{xi}}$$

and

$$\sigma\{X1 \cup ... \cup XM\} = \sqrt{\frac{\sum_{i=1}^{M} N_{Xi}\left(\sigma_{Xi}^2 + \mu_{Xi}^2\right)}{\sum_{i=1}^{M} N_{xi}} - \mu^2\{X1 \cup ... \cup XM\}}$$

where

$$X_i \cap Xj \equiv \theta$$
$$\forall_i \neq j$$

If the size (actual or relative to one another), mean, and standard deviation of two overlapping populations are known for the populations as well as their intersection, then the standard deviation of the overall population can still be calculated as follows:

$$\mu_{X \cup Y} = \frac{N_X \mu_X + N_Y \mu_Y - N_{X \cap Y} \mu_{X \cap Y}}{N_X + N_Y - N_{X \cap Y}}$$

and

$$\sigma_{X \cup Y} =$$

$$\sqrt{\frac{N_X\left(\sigma_X^2 + \mu_X^2\right) + N_Y\left(\sigma_Y^2 + \mu_Y^2\right) - N_{X \cup Y}\left(\sigma_{X \cap Y}^2 + \mu_{X \cap Y}^2\right)}{N_X + N_Y - N_{X \cap Y}} - \mu_{X \cup Y}^2}$$

If two or more sets of data are being added in a pairwise fashion, the standard deviation can be calculated if the covariance between the each pair of data sets is known.

$$\sigma_{X1+...+XM} \sqrt{\sum_{i=1}^{M}\left(\sigma_{Xi}^2\right) - \sum_{i=1}^{M}\sum_{j=1}^{M} Cov(Xi, Xj)}$$

For the special case where no correlation exists between all pairs of data sets, then the relation reduces to:

$$\sigma_{X1+...+XM} \sqrt{\sum_{i=1}^{M}\left(\sigma_{Xi}^2\right)}$$

where,

$$\mathrm{Cov}(Xi, xj) = 0$$
$$\forall\{i, j\}$$

Sample-based Statistics

Standard deviations of non-overlapping sub-samples can be aggregated as follows if the actual size and means of each are known:

$$\mu_{X\cup Y} = \frac{N_X\mu_X + N_Y\mu_Y}{N_X + N_Y}$$

and:

$$\sigma_{X\cup Y} = \sqrt{\frac{(N_{X-1})\sigma_X^2 + N_X\mu_X^2 + (N_Y - 1)\sigma_Y^2 + N_Y\mu_Y^2 - (N_X + N_Y)\mu_{X\cup Y}^2}{N_X + N_Y - 1}}$$

where,

$$X \cap Y \equiv \theta$$

For the more general M non-overlapping data sets X1 through XM:

$$\mu\{x1 \cup \ldots \cup XM\} = \frac{\sum_{i=1}^{M} N_{Xi}\mu_{Xi}}{\sum_{i=1}^{M} N_{Xi}}$$

and:

$$\mu\{X1 \cup \ldots \cup XM\} = \frac{\sum_{i=1}^{M}\left((N_{Xi} - 1)\sigma_{Xi}^2 + N_{Xi}\mu_{Xi}^2\right) - \left(\sum_{i=1}^{M} N_{Xi}\right)\mu_{\{X1\cup \ldots \cup XM\}}^2}{\sum_{i=1}^{M} N_{Xi} - 1}$$

where,

$$Xi \cap Xj \equiv \theta$$
$$\forall i \neq j$$

If the size, mean, and standard deviation of two overlapping samples are known for the samples as well as their intersection, then the standard deviation of the samples can still be calculated. In general:

$$\mu_{X\cup Y} = \frac{N_X\mu_X + N_Y\mu_Y - N_{X\cap Y}\mu_{X\cap Y}}{N_X + N_Y - N_{X\cap Y}}$$

and:

$$\sigma_{X\cup Y} = \sqrt{\frac{(N_x - 1)\sigma_x^2 + N_X\mu_X^2 + (N_Y - 1)\sigma_Y^2 - (N_{X\cap Y} - 1)\sigma_{X\cap Y}^2 - N_{X\cap Y}\mu_{X\cap Y}^2 - (N_X + N_Y - N_{X\cap Y})\mu_{X\cup Y}^2}{N_X + N_Y - N_{X\cap Y} - 1}}$$

COEFFICIENT OF VARIATION

In probability theory and statistics, the coefficient of variation (CV) is a normalized measure of dispersion of a probability distribution. It is defined as the ratio of the standard deviation σ to the mean μ:

$$C_v = \frac{\sigma}{\mu}$$

This is only defined for non-zero mean, and is most useful for variables that are always positive. It is also known as unitized risk or the variation coefficient. It is expressed as percentage. The coefficient of variation should only be computed for data measured on a ratio scale.

As an example, if a group of temperatures are Analysed, the standard deviation does not depend on whether the Kelvin or Celsius scale is used since an object that changes its temperature by 1 K also changes its temperature by 1 C. However the mean temperature of the data set would differ in each scale by an amount of 273 and thus the coefficient of variation would differ. So the coefficient of variation does not have any meaning for data on an interval scale.

Standardized moments are similar ratios,

$$\frac{\mu k}{\sigma^k},$$

which are also dimensionless and scale invariant. The variance-to-mean ratio, σ^2/μ, is another similar ratio, but is not dimensionless, and hence not scale invariant.

In signal processing, particularly image processing, the reciprocal ratio μ/σ is referred to as the signal to noise ratio.

COMPARISON TO STANDARD DEVIATION

Advantages

The coefficient of variation is useful because the standard

deviation of data must always be understood in the context of the mean of the data. The coefficient of variation is a dimensionless number.

So when comparing between data sets with different units or widely different means, one should use the coefficient of variation for comparison instead of the standard deviation.

Disadvantages

- When the mean value is near zero, the coefficient of variation is sensitive to small changes in the mean, limiting its usefulness.
- Unlike the standard deviation, it cannot be used to construct confidence intervals for the mean.

APPLICATIONS

The coefficient of variation is also common in applied probability fields such as renewal theory, queueing theory, and reliability theory.

In these fields, the exponential distribution is often more important than the normal distribution. The standard deviation of an exponential distribution is equal to its mean, so its coefficient of variation is equal to 1.

Distributions with CV < 1 (such as an Erlang distribution) are considered low-variance, while those with CV > 1 (such as a hyper-exponential distribution) are considered high-variance.

Some formulas in these fields are expressed using the squared coefficient of variation, often abbreviated SCV. In modeling, a variation of the CV is the CV(RMSD). Essentially the CV(RMSD) replaces the standard deviation term with the Root Mean Square Deviation (RMSD).

DISTRIBUTION

Under weak conditions on the sample distribution, the probability distribution of the coefficient of variation is known. In fact, it has been determined by Hendricks and Robey. This is useful, for instance, in the construction of hypothesis tests or confidence intervals.

SKEWNESS AND KURTOSIS

A fundamental task in many statistical analyses is to characterize the location and variability of a data set. A further characterization of the data includes skewness and kurtosis.

Skewness is a measure of symmetry, or more precisely, the lack of symmetry. A distribution, or data set, is symmetric if it looks the same to the left and right of the centre point.

Kurtosis is a measure of whether the data are peaked or flat relative to a normal distribution. That is, data sets with high kurtosis tend to have a distinct peak near the mean, decline rather rapidly, and have heavy tails. Data sets with low kurtosis tend to have a flat top near the mean rather than a sharp peak. A uniform distribution would be the extreme case.

The histogram is an effective graphical technique for showing both the skewness and kurtosis of data set.

DEFINITION OF SKEWNESS

For univariate data Y_1, Y_2..., Y_N, the formula for skewness is:

$$\text{skewness} = \frac{\sum_{i=1}^{N}\left(Y_i - +\bar{Y}\right)^3}{(N-1)s^3}$$

where $\bar{Y}$ is the mean, s is the standard deviation, and N is the number of data points. The skewness for a normal distribution is zero, and any symmetric data should have a skewness near zero.

Negative values for the skewness indicate data that are skewed left and positive values for the skewness indicate data that are skewed right.

By skewed left, we mean that the left tail is long relative to the right tail. Similarly, skewed right means that the right tail is long relative to the left tail. Some measurements have a lower bound and are skewed right. For example, in reliability studies, failure times cannot be negative.

DEFINITION OF KURTOSIS

For univariate data Y_1, Y_2..., Y_N, the formula for kurtosis is:

$$\text{kurtosis} = \frac{\sum_{i=1}^{N}\left(Y_i - + \bar{Y}\right)^4}{(N-1)s^4}$$

where $\bar{Y}$ is the mean, s is the standard deviation, and N is the number of data points.

DEALING WITH SKEWNESS AND KURTOSIS

Many classical statistical tests and intervals depend on normality assumptions. Significant skewness and kurtosis clearly indicate that data are not normal. If a data set significant skewness or kurtosis (as indicated by a histogram or the numerical measures), what can we do about it?

One approach is to apply some type of transformation to try to make the data normal, or more nearly normal. The Box-Cox transformation is a useful technique for trying to normalize a data set. In particular, taking the log or square root of a data set is often useful for data that moderate right skewness.

Another approach is to use techniques based on distributions other than the normal. For example, in reliability studies, the exponential, Weibull, and lognormal distributions are typically used as a basis for modeling rather than using the normal distribution. The probability plot correlation coefficient plot and the probability plot are useful tools for determining a good distributional model for the data.

INDEX NUMBER

An index number is an economic data figure reflecting price or quantity compared with a standard or base value. The base usually equals 100 and the index number is usually expressed as 100 times the ratio to the base value.

For example, if a commodity costs twice as much in 1970 as it did in 1960, its index number would be 200 relative to 1960. Index numbers are used especially to compare business activity, the cost of living, and employment. They enable economists to reduce unwieldy business data into easily understood terms.

In economics, index numbers generally are time series summarising movements in a group of related variables. In some cases, however, index numbers may compare geographic areas at a point in time. An example is a country's purchasing power parity. The best-known index number is the consumer price index, which measures changes in retail prices paid by consumers. In addition, a cost-of-living index (COLI) is a price index number that measures relative cost of living over time. In contrast to a COLI based on the true but unknown utility function, a superlative index number is an index number that can be calculated. Thus, superlative index numbers are used to provide a fairly close approximation to the underlying cost-of-living index number in a wide range of circumstances.

There is a substantial body of economic analysis concerning the construction of index numbers, desirable properties of index numbers and the relationship between index numbers and economic theory.

4

Correlation Analysis

RANKING METHOD

The ranking system is a non-quantitative method of listing jobs in order of the demands they are considered to make on those who perform them or their importance to the unit. The evaluators are required to keep the whole job in mind and to compare the difficulty and requirements of each job with every other job in the unit. This is usually done either by sorting the job descriptions (or an abbreviation form of the job description such as a job specification) on the basis of difficulty/importance into a rank order from most difficult/importance to least difficult/importance, or by the method of paired comparisons. In the later method, each job is compared with every other job being evaluated by making separate judgements between every pair of jobs.

It is particularly important in the ranking method that evaluators are familiar with all the jobs being ranked and that they have full information about the jobs, preferably in the form of complete job descriptions. After the jobs have been listed in order of difficulty the evaluators, or probably a central committee in a large organization, determine the number of levels to be established for pay purposes and the groupings of the jobs into levels.

CORRELATION COEFFICIENT

The correlation coefficient, sometimes also called the cross-correlation coefficient, is a quantity that gives the quality of a least squares fitting to the original data. To define the correlation coefficient, first consider the sum of squared values SS_{xx}, SS_{xy}, and SS_{yy} of a set of n data points (x_i, y_i) about their respective means,

$$SS_{xx} \equiv \sum(x_i - \bar{x})^2$$
$$= \sum x^2 - 2\bar{x}\sum x + \sum \bar{x}^2$$
$$= \sum x^2 - 2n\bar{x}^2 + n\bar{x}^2$$
$$= \sum x^2 - n\bar{x}^2$$
$$SS_{yy} \equiv \sum(y_i - \bar{y})^2$$
$$= \sum y^2 - 2\bar{y}\sum y + \sum \bar{y}^2$$
$$= \sum y^2 - 2n\bar{y}^2 + n\bar{y}^2$$
$$= \sum y^2 - 2n\bar{y}^2 + n\bar{y}^2$$
$$= \sum y^2 - n\bar{y}^2$$
$$SS_{xy} \equiv \sum y^2 - n\bar{y}^2$$
$$= \sum(x_i - \bar{x})(y_i - \bar{y})$$
$$= \sum(x_i y_i - \bar{x} y_i - x_i \bar{y} + \bar{x}\bar{y})$$
$$= \sum xy - n\bar{x}\bar{y} - n\bar{x}\bar{y} + n\bar{x}\bar{y}$$
$$= \sum xy - n\bar{x}\bar{y}$$

These quantities are simply unnormalized forms of the variances and covariance of Xand Y given by,

$$S_{xx} = N\,\mathrm{var}(X)$$
$$SS_{yy} = N\,\mathrm{var}(Y)$$
$$SS_{xy} = N\,\mathrm{cov}(X, Y).$$

For linear least squares fitting, the coefficient b in,

$$y = a + bx$$

is given by,

$$b = \frac{n\sum xy - \sum x \sum y}{n\sum x^2 - \left(\sum x\right)^2}$$
$$= \frac{SS_{xy}}{SS_{xx}},$$

and the coefficient b′ in,

$$x = a' + b'y$$

is given by,

$$b' = \frac{n\sum xy - \sum x \sum y}{n\sum y^2 - \left(\sum y\right)^2}$$

The correlation coefficient r (sometimes also denoted R) is then defined by,

$$r^2 \equiv bb'$$

$$= \frac{SS_{xy}^2}{SS_{xx}SS_{yy}}$$

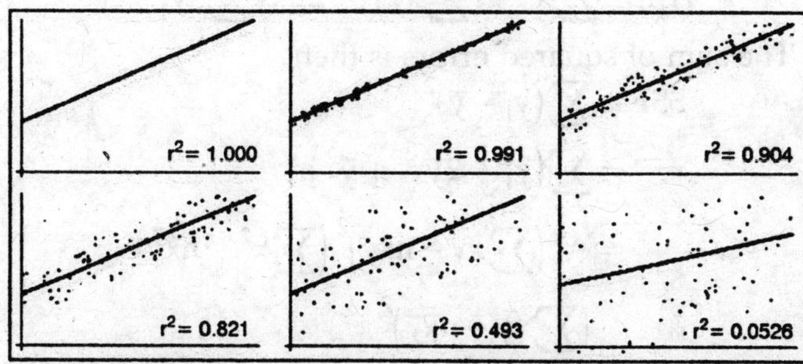

The correlation coefficient is also known as the product-moment coefficient of correlation or Pearson's correlation. The correlation coefficients for linear fits to increasingly noisy data are shown.

The correlation coefficient has an important physical interpretation.

$$A \equiv \left[\sum x^2 - n\bar{x}^2\right]^{-1}$$

and denote the"expected" value for y_i as $\hat{y}_i$. Sums of $\hat{y}_i$ are then

$$\begin{aligned}
\hat{y}_i &= a + bx_i \\
&= \bar{y} - b\bar{x} + bx_i \\
&= \bar{y} + b(x_i - \bar{x}) \\
&= A\left(\bar{y}\sum x^2 - \bar{x}\sum xy + x_i\sum xy - n\bar{x}\bar{y}x_i\right) \\
&= A\left[\bar{y}\sum x^2 + (x_i - \bar{x})\sum xy - n\bar{x}\bar{y}x_i\right]
\end{aligned}$$

$$\sum \hat{y}_i = A\left(n\overline{y}\sum x^2 - n^2\overline{x}^2\overline{y}\right)$$

$$\sum \hat{y}_i^2 =$$

$$A^2\left[\begin{aligned}&n\overline{y}^2\left(\sum x^2\right)^2 - n^2\overline{x}^2\overline{y}^2\left(\sum x^2\right)\\&-2n\overline{x}\,\overline{y}\left(\sum xy\right)\left(\sum x^2\right)\left(\sum xy\right)^2 - n\overline{x}^2\left(\sum xy\right)\end{aligned}\right]$$

$$\sum y_i\hat{y}_i = A\sum\left[y_i\overline{y}\sum x^2 + y_i\left(x_i - \overline{x}\right)\sum xy - n\overline{xy}x_iy_i\right]$$

$$= A\left[n\overline{y}^2\sum x^2 + \left(\sum xy\right)^2 - n\overline{xy}\sum xy - n\overline{xy}\left(\sum xy\right)\right]$$

$$= A\left[n\overline{y}^2\sum x^2 + \left(\sum xy\right)^2 - 2n\overline{xy}\sum xy\right]$$

The sum of squared errors is then,

$$\text{SSE} \equiv \sum\left(\hat{y}_i - \overline{y}\right)^2$$

$$= \sum\left(\hat{y}_i^2 - 2\overline{y}\hat{y}_i + \overline{y}^2\right)$$

$$= A^2\left(\sum xy - n\overline{xy}\right)^2\left(\sum x^2 - n\overline{x}^2\right)$$

$$= \frac{\left(\sum xy - n\overline{xy}\right)^2}{\sum x^2 - n\overline{x}^2}$$

$$= b\,SS_{xy}$$

$$= \frac{SS_{xy}^2}{SS_{xx}}$$

$$= SS_{yy}r^2$$

$$= b^2SS_{xx}.$$

and the sum of squared residuals is

$$\text{SSR} \equiv \sum\left(y_i - \hat{y}_i\right)^2$$

$$= \sum\left(y_i - \overline{y} + b\overline{x} - bx_i\right)^2$$

$$= \sum\left[y_i - \overline{y} - b\left(x_i - \overline{x}\right)\right]^2$$

$$= \sum\left(y_i - \overline{y}\right)^2 + b^2\sum\left(x_i - \overline{x}\right)^2 - 2b\sum\left(x_i - \overline{x}\right)\left(y_i - \overline{y}\right)$$

$$= SS_{yy} + b^2SS_{xx} - 2bSS_{xy}.$$

But,

$$b = \frac{SS_{xy}}{SS_{xy}}$$

$$r^2 = \frac{SS_{xy}^2}{SS_{xx}SS_{yy}}$$

so

$$SSE = SS_{yy} + \frac{SS_{xy}^2}{SS_{xx}^2} SS_{xx} - 2\frac{SS_{xy}}{SS_{xx}} SS_{xy}$$

$$= SS_{yy} - \frac{SS_{xy}^2}{SS_{xx}}$$

$$= SS_{yy}\left(1 - \frac{SS_{xy}^2}{SS_{xx}SS_{yy}}\right)$$

$$= SS_{yy}\left(1 - r^2\right).$$

and

$$SSE + SSR = SS_{yy}\left(1 - r^2\right) + SS_{yy}r^2 = SS_{yy}.$$

The square of the correlation coefficient r^2 is therefore given by,

$$r^2 \equiv \frac{SSR}{SS_{yy}}$$

$$= \frac{SS_{xy}^2}{S_{xx}SS_{yy}}$$

$$= \frac{\left(\sum xy - n\bar{x}\bar{y}\right)^2}{\left(\sum x^2 - n\bar{x}^2\right)\left(\sum y^2 - n\bar{y}^2\right)}$$

In other words, r^2 is the proportion of SS_{yy} which is accounted for by the regression. If there is complete correlation, then the lines obtained by solving for best-fit (a, b)and (a′, b′)coincide (since all data points lie on them), so solving () for y and equating to () gives,

$$y = -\frac{a'}{b'} + \frac{x}{b'} = a + bx$$

Therefore, $a = -a'/b'$ and $b = -1/b'$, giving

$$r^2 = bb' = 1$$

The correlation coefficient is independent of both origin and scale, so

$$r(u,v) = r(x,y)$$

where,

$$u \equiv \frac{x - x_0}{h}$$

$$v \equiv \frac{y - y_0}{h}$$

PROPERTIES OF CORRELATION

- Correlation requires that both variables be quantitative (numerical).
  You can't calculate a correlation between"income" and"city of residence"because"city of residence" is a qualitative (non-numerical) variable.
- Positive r indicates positive association between the variables, and negative r indicates negative association.
  A positive r indicates that above average values of x tend to be matched with above average values of y and below average values of x tend to be matched with below average values of y.
  Positive r high with high, low with low A negative r indicates that above average values of x tend to be matched with below average values of y and below average values of x tend to be matched with above average values of y.
  Negative r high with low, low with high
- The correlation coefficient (r) is always a number between -1 and +1.
  Values of r near 0 indicate a very weak linear relationship. The extreme values of -1 and +1 indicate the points in a scatterplot lie exactly along a straight line.
- The correlation coefficient (r) is a pure number without units.

r is not affected by:

- Interchanging the two variables (it makes no difference which variable is called x and which is called y)
- Adding the same number to all the values of one variable
- Multiplying all the values of one variable by the same positive number

Because r uses the standardized values of the observations, r does not change when we change units of measurement (inches vs. centimetres, pounds vs. kilograms, miles vs. metres). r is"scale invariant".

REGRESSION ANALYSIS

He goal of regression analysis is to determine the values of parametres for a function that cause the function to best fit a set of data observations that you provide.

In linear regression, the function is a linear (straight-line) equation. For example, if we assume the value of an automobile decreases by a constant amount each year after its purchase, and for each mile it is driven, the following linear function would predict its value (the dependent variable on the left side of the equal sign) as a function of the two independent variables which are age and miles:

value = price + depage*age + depmiles*miles

where value, the dependent variable, is the value of the car, age is the age of the car, and miles is the number of miles that the car has been driven.

The regression analysis performed by NLREG will determine the best values of the three parametres, price, the estimated value when age is 0 (i.e., when the car was new), depage, the depreciation that takes place each year, and depmiles, the depreciation for each mile driven. The values of depage and depmiles will be negative because the car loses value as age and miles increase.

For an analysis such as this car depreciation example, you must provide a data file containing the values of the dependent and independent variables for a set of observations.

In this example each observation data record would contain three numbers: value, age, and miles, collected from used car ads for the same model car.

The more observations you provide, the more accurate will be the estimate of the parametres. The NLREG statements to perform this regression:

```
Variables value,age,miles;
Parametres price,depage,depmiles;
Function value = price + depage*age +
depmiles*miles;
Data;
{data values go here}
```

Once the values of the parametres are determined by NLREG, you can use the formula to predict the value of a car based on its age and miles driven. For example, if NLREG computed a value of 16000 for price, -1000 for depage, and -0.15 for depmiles, then the function

```
value = 16000 - 1000*age - 0.15*miles
```

could be used to estimate the value of a car with a known age and number of miles.

If a perfect fit existed between the function and the actual data, the actual value of each car in your data file would exactly equal the predicted value.

Typically, however, this is not the case, and the difference between the actual value of the dependent variable and its predicted value for a particular observation is the error of the estimate which is known as the"deviation" or"residual". The goal of regression analysis is to determine the values of the parametres that minimize the sum of the squared residual values for the set of observations. This is known as a"least squares" regression fit.

Here is a plot of a linear function fitted to a set of data values. The actual data points are marked with"x". The red line between a point and the fitted line represents the residual for the observation. NLREG is a very powerful regression analysis programme. Using it you can perform multivariate, linear, polynomial, exponential, logistic, and general nonlinear regression.

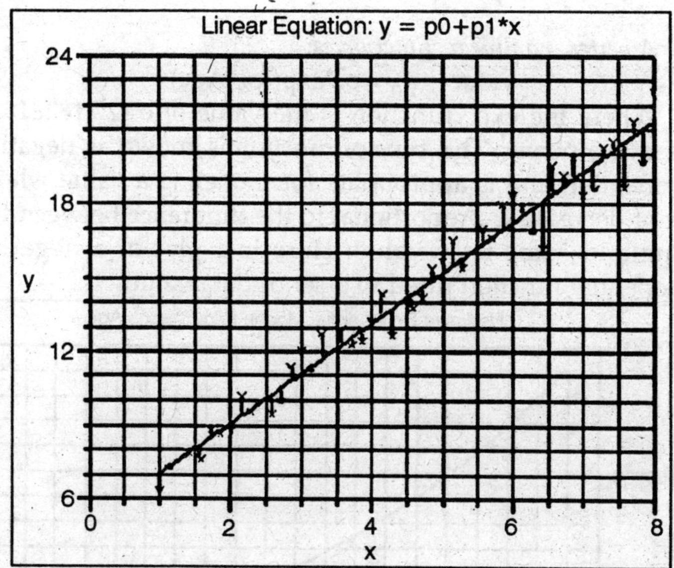

What this means is that you specify the form of the function to be fitted to the data, and the function may include nonlinear terms such as variables raised to powers and library functions such as log, exponential, sine, etc. For complex analyses, NLREG allows you to specify function models using conditional statements (if, else), looping (for, do, while), work variables, and arrays. NLREG uses a state-of-the-art regression algorithm that works as well, or better, than any you are likely to find in any other, more expensive, commercial statistical packages. As an example of nonlinear regression, consider another depreciation problem. The value of a used airplane decreases for each year of its age. Assuming the value of a plane falls by the same amount each year, a linear function relating value to age is:

$$value = p_0 + p_1{*}Age$$

Where p_0 and p_1 are the parametres whose values are to be determined. However, it is a well-known fact that planes (and automobiles) lose more value the first year than the second, and more the second than the third, etc. This means that a linear (straight-line) function cannot accurately model this situation.

A better, nonlinear, function is:

$$value = p_0 + p_1 * exp(-p_2 * Age)$$

Where the"exp" function is the value of e (2.7182818...) raised to a power. This type of function is known as"negative exponential" and is appropriate for modeling a value whose rate of decrease is proportional to the difference between the value and some base value. Here is a plot of a negative exponential function fitted to a set of data values.

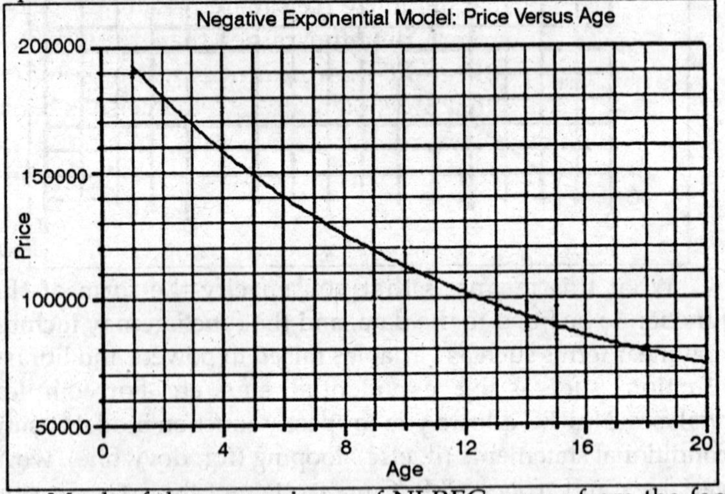

Much of the convenience of NLREG comes from the fact that you can enter complicated functions using ordinary algebraic notation.

Examples of functions that can be handled with NLREG include:

- *Linear:* Y = p0 + p1*X
- *Quadratic:* Y = p0 + p1*X + p2*X^2
- *Multivariate:* Y = p0 + p1*X + p2*Z + p3*X*Z
- *Exponential:* Y = p0 + p1*exp(X)
- *Periodic:* Y = p0 + p1*sin(p2*X)
- *Misc:* Y = p0 + p1*Y + p2*exp(Y) + p3*sin(Z)

In other words, the function is a general expression involving one dependent variable (on the left of the equal sign), one or more independent variables, and one or more parametres whose values are to be estimated. NLREG can handle up to 500 variables and 500 parametres. Because of its generality, NLREG can perform all of the regressions handled by ordinary linear or multivariate regression programmes as well as nonlinear regression.

Some other regression programmes claim to perform nonlinear regression but actually do it by transforming the values of the variables such that the function is converted to linear form. They then perform a linear regression on the transformed function.

This technique has a major flaw: it determines the values of the parametres that minimize the squared residuals for the transformed, linearized function rather than the original function. This is different than minimizing the squared residuals for the actual function and the estimated values of the parametres may not produce the best fit of the original function to the data. NLREG uses a true nonlinear regression technique that minimizes the squared residuals for the actual function.

FITTING LINEAR DATA WITH NONLINEAR REGRESSION

USING NONLINEAR REGRESSION TO ACCESS OPTIONS

Nonlinear regression is more versatile than linear regression. Since nonlinear regression can fit data to any model, even a linear one, you may find it useful occasionally to use a nonlinear regression analysis to Analyse linear data. This lets you take advantage of the increased flexibility of many nonlinear regression programmes.

For example, you could use the nonlinear regression analysis to compare an unconstrained linear regression line with one forced through the origin (or some other point). Or you could fit a linear regression line, but weight the points to minimize the sum of the relative distance squared, rather than the distance squared.

DETERMINING A POINT OTHER THAN THE Y INTERCEPT

The linear regression equation defines Y as a function of slope and Y intercept:

$$y = \text{intercept} + \text{slope} \cdot X$$

This equation can be rewritten in a more general form, where we replace the intercept (Y value at X=0) with YX', the Y value at X', where X' is some specified X value. Now the linear regression equation becomes:

$$Y = Y_{X'} + \text{slope} \cdot (x - x')$$

Y at any particular X value equals the Y value at X' plus the slope times the distance from X to X'. If you set X' to zero, this equation becomes identical to the previous one.

This equation has three variables, YX', X', and slope. If you set either X' or YX' to a constant value, you can fit the other (and the slope).

When you fit these equations using nonlinear regression, Prism insists that you enter rules for initial values (or directly enter initial values). Since the equation is in fact linear, Prism will find the best-fit values no matter what initial values you enter. Setting all initial values to zero should work fine.

Note. When you enter your equation, you must define either X' or YX'. The two are related, so it is impossible to fit both.

Example: You want to find the X value (with SE and confidence interval) where Y=50 for linear data. Fit the data using nonlinear regression using this equation to determine the best-fit values of the slope and X50.

Y = 50 + slope*(X–X50)

In this equation, X50 is the value of X where Y=50. Don't confuse this with an EC_{50}, which is the value of X where Y is halfway between minimum and maximum in a dose-response curve. A linear model does not have a minimum and maximum, so the concept of EC50 does not apply.

Example: You want to fit data to a straight line to determine the slope and the Y value at X=20. Fit the data using nonlinear regression to this equation.

Y = Y20 + slope*(X–20)

Example: You want to determine the slope and the X intercept, with SE and confidence interval. Fit the data using nonlinear regression to this equation:

Y = slope*(X–Xintercept)

Fitting Two Line Segments

Nonlinear regression can fit two line segments to different portions of the data, so that the two meet at a defined X value.

FITTING STRAIGHT LINES TO SEMI-LOG GRAPHS

You can start linear (or nonlinear) regression from a data table, results table or graph. Prism then fits a model to the data. If you start from a graph, Prism fits to the data plotted on that graph. Selecting a logarithmic axis (from the Axis dialog) does not change the data, so does not change the way Prism performs regression. If you plot a linear regression"line" on a graph with a logarithmic axis, the best-fit"line" will be curved. To fit a straight line on a semilog plot requires use of nonlinear regression.

If the Y-axis is logarithmic use this equation.

Y=10^(Slope*X + Yintercept)

Graphing on a log Y-axis is equivalent to taking the antilog. The antilog of the right side of the equation is the equation for a straight line: slope*X + Yintercept. So this equation appears linear when graphed on a logarithmic Y-axis. It is difficult to define rules for initial values that work for all data sets, so you'll need to enter initial values individually for each data set. Here is an example.

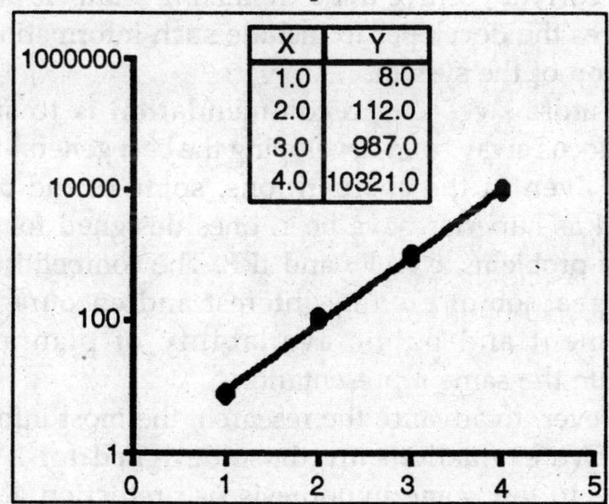

X	Y
1.0	8.0
2.0	112.0
3.0	987.0
4.0	10321.0

If the X-axis is logarithmic, then use this equation:

Y = Slope*10^X + Yintercept

If both axes are logarithmic (rare) then use this equation:

Y = 10^(Slope*10^X + Yintercept)

INTERPRETATION OF RESULTS AND RECOMMENDATIONS

We reflect on what those results mean for empirical comparison of planners; we summarize the results and recommend some partial solutions. It is not possible to guarantee fairness and we propose no magic formula for performing evaluations, but the state of the practice in general can certainly be improved. We propose three general recommendations and 12 recommendations targeted to specific assumptions.

Many of the targeted recommendations amount to requesting problem and planner developers to be more precise about the requirements for and expectations of their contributions. Because the planners are extremely complex and time consuming to build, the documentation may be inadequate to determine how a subsequent version differs from the previous or under what conditions (e.g., parametre settings, problem types) the planner can be fairly compared. With the current positive trend in making planners available, it behooves the developer to include such information in the distribution of the system.

The most sweeping recommendation is to shift the research focus away from developing the best general-purpose planner. Even in the competitions, some of the planners identified as superior have been ones designed for specific classes of problems, e.g., FF and IPP. The competitions have done a great job of exciting interest and encouraging the development and public availability of planners that incorporate the same representation.

However, to advance the research, the most informative comparative evaluations are those designed for a specific purpose - to test some hypothesis or prediction about the performance of a planner. An experimental hypothesis focuses the analysis and often leads naturally to justified design decisions about the experiment itself. For example, Hoffmann and Nebel, the authors of the Fast-Forward (FF) system, state in the introduction to their JAIR paper that FF's development was motivated by a specific set of the benchmark domains; because the system is heuristic, they designed the heuristics to fit the expectations/needs of those domains. Additionally, in part of their evaluation, they compare to a specific system on which their own system had commonalities and point out the various advantages or disadvantages of their design decisions on specific problems. Follow-up work or researchers comparing their own systems to FF now have a well-defined starting point for any comparison.

Recommendation: Experiments should be driven by hypotheses. Researchers should precisely articulate in advance of the experiments their expectations about how their new planner or augmentations to an existing planner add to the state of the art. These expectations should in turn justify the selection of problems, other planners and metrics that form the core of the comparative evaluation.

A general issue is whether the results are accurate. We reported the results as they are output by the planners. If a planner stated in its output that it had been successful, we took it at face value.

However, by examining some of the output, we determined that some claims of successful solution were erroneous - the proposed solution would not work. The only way to ensure that the output is correct is with a solution checker.

Drew McDermott used a solution checker in the AIPS98 competition. However, the planners do not all provide output in a compatible format with his checker. Thus, another concern with any comparative evaluation is that the output needs to be cross-checked. Because we are not declaring a winner, we do not think that the lack of a solution checker casts serious doubt on our results. For the most part, we have only been concerned with factors that cause the observed success rates to change.

Recommendation: Just as input has been standardized with PDDL, output should be standardized, at least in the format of returned plans.

Another general issue is whether the benchmark sets are representative of the space of interesting planning problems. We did not test this directly (in fact, we are not sure how one could do so), but the clustering of results and observations by others in the planning community suggest that the set is biased towards logistics problems.

Additionally, many of the problems are getting dated and no longer distinguish performance. Some researchers have begun to more formally Analyse the problem set, either in service of building improved planners or to better understand planning problems.

For example, in the related area of scheduling, our group has identified distinctive patterns in the topology of search spaces for different types of classical scheduling problems and has related the topology to performance of algorithms. Within planning, Hoffman has examined the topology of local search spaces in some of the small problems in the benchmark collection and found a simple structure with respect to some well-known relaxations.

Additionally, he has worked out a partial taxonomy, based on three characteristics, for the Analysed domains. Helmert has Analysed the computational complexity of a subclass of the benchmarks, transportation problems, and has identified key features that affect the difficulty of such problems.

Recommendation: The benchmark problem sets should themselves be evaluated and over-hauled. Problems that can be easily solved should be removed. Researchers should study the benchmark problems/domains to classify them into problem types and key characteristics. Developers should contribute application problems and realistic versions of them to the evolving set.

The remainder of this part describes other recommendations for improving the state of the art in planner comparisons.

RELATIONSHIP BETWEEN REGRESSION AND CORRELATION

AN OVERVIEW

Regression analysis involves identifying the relationship between a dependent variable and one or more independent variables. A model of the relationship is hypothesized, and estimates of the parametre values are used to develop an estimated regression equation. Various tests are then employed to determine if the model is satisfactory. If the model is deemed satisfactory, the estimated regression equation can be used to predict the value of the dependent variable given values for the independent variables.

Regression Model

In simple linear regression, the model used to describe the relationship between a single dependent variable y and a single independent variable x is y = a0 + a1x + k. a0and a1 are referred to as the model parametres, and is a probabilistic error term that accounts for the variability in y that cannot be explained by the linear relationship with x. If the error term were not present, the model would be deterministic; in that case, knowledge of the value of x would be sufficient to determine the value of y.

Least Squares Method

Either a simple or multiple regression model is initially posed as a hypothesis concerning the relationship among the dependent and independent variables. The least squares method is the most widely used procedure for developing estimates of the model parametres.

As an illustration of regression analysis and the least squares method, suppose a university medical centre is investigating the relationship between stress and blood pressure. Assume that both a stress test score and a blood pressure reading have been recorded for a sample of 20 patients. The data are shown graphically in the figure, called a scatter diagram. Values of the independent variable, stress test score, are given on the horizontal axis, and values of the dependent variable, blood pressure, are shown on the vertical axis. The line passing through the data points is the graph of the estimated regression equation: y = 42.3 + 0.49x. The parametre estimates, b0 = 42.3 and b1 = 0.49, were obtained using the least squares method.

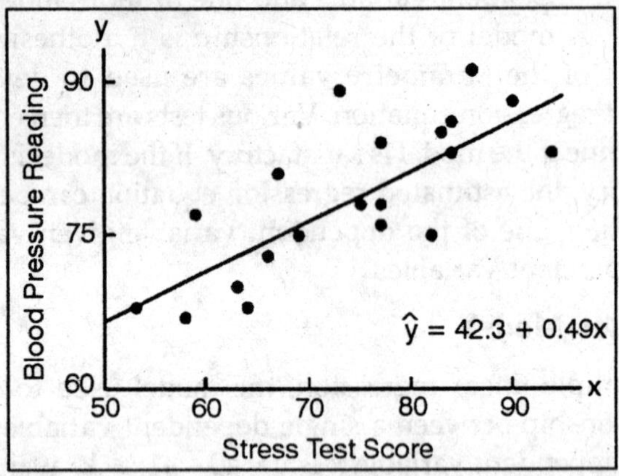

Correlation

Correlation and regression analysis are related in the sense that both deal with relationships among variables. The correlation coefficient is a measure of linear association between two variables. Values of the correlation coefficient are always between –1 and +1. A correlation coefficient of +1 indicates that two variables are perfectly related in a positive linear sense, a correlation coefficient of –1 indicates that two variables are perfectly related in a negative linear sense, and a correlation coefficient of 0 indicates that there is no linear relationship between the two variables. For simple linear regression, the sample correlation coefficient is the square root of the coefficient of determination, with the sign of the correlation coefficient being the same as the sign of b1, the coefficient of x1 in the estimated regression equation.

Neither regression nor correlation analyses can be interpreted as establishing cause-and-effect relationships. They can indicate only how or to what extent variables are associated with each other. The correlation coefficient measures only the degree of linear association between two variables. Any ceases about a cause-and-effect relationship must be based on the judgment of the analyst.

TIME SERIES ANALYSIS

Time series data often arise when monitoring industrial processes or tracking corporate business metrics. The essential difference between modeling data via time series methods or using the process monitoring methods discussed is the following:

Time series analysis accounts for the fact that data points taken over time may have an internal structure (such as autocorrelation, trend or seasonal variation) that should be accounted for.

DEFINITIONS, APPLICATIONS AND TECHNIQUES

Definition of Time Series

An ordered sequence of values of a variable at equally spaced time intervals.

Applications

The usage of time series models is twofold:

- Obtain an understanding of the underlying forces and structure that produced the observed data
- Fit a model and proceed to forecasting, monitoring or even feedback and feedforward control.

Time Series Analysis is used for many applications such as:

- Economic Forecasting
- Sales Forecasting
- Budgetary Analysis
- Stock Market Analysis
- Yield Projections
- Process and Quality Control
- Inventory Studies
- Workload Projections
- Utility Studies
- Census Analysis

Techniques

The fitting of time series models can be an ambitious undertaking. There are many methods of model fitting including the following:

- Box-Jenkins ARIMA models
- Box-Jenkins Multivariate Models
- Holt-Winters Exponential Smoothing (single, double, triple)

The user's application and preference will decide the selection of the appropriate technique. The overview presented here will start by looking at some basic smoothing techniques:

- Averaging Methods
- Exponential Smoothing Techniques.

As suggested, discuss the Box-Jenkins modeling methods and Multivariate Time Series.

ANALYSIS

There are several types of data analysis available for time series which are appropriate for different purposes.

General Exploration

- Graphical examination of data series
- Autocorrelation analysis to examine serial dependence
- Spectral analysis to examine cyclic behaviour which need not be related to seasonality. For example, sun spot activity varies over 11 year cycles. Other common examples include celestial phenomena, weather patterns, commodity prices, and economic activity.

Description

- Separation into components representing trend, seasonality, slow and fast variation, cyclical irregular:
- Simple properties of marginal distributions

Prediction and Forecasting

- Fully-formed statistical models for stochastic

simulation purposes, so as to generate alternative versions of the time series, representing what might happen over non-specific time-periods in the future (prediction).

- Simple or fully-formed statistical models to describe the likely outcome of the time series in the immediate future, given knowledge of the most recent outcomes (forecasting).

MODELS

Models for time series data can have many forms and represent different stochastic processes. When modeling variations in the level of a process, three broad classes of practical importance are the autoregressive (AR) models, the integrated (I) models, and the moving average (MA) models.

These three classes depend linearly on previous data points. Combinations of these ideas produce autoregressive moving average (ARMA) and autoregressive integrated moving average (ARIMA) models. The autoregressive fractionally integrated moving average (ARFIMA) model generalizes the former three. Extensions of these classes to deal with vector-valued data are available under the heading of multivariate time-series models and sometimes the preceding acronyms are extended by including an initial"V" for"vector".

An additional set of extensions of these models is available for use where the observed time-series is driven by some"forcing" time-series (which may not have a causal effect on the observed series): the distinction from the multivariate case is that the forcing series may be deterministic or under the experimenter's control. For these models, the acronyms are extended with a final"X" for"exogenous".

Non-linear dependence of the level of a series on previous data points is of interest, partly because of the possibility of producing a chaotic time series. However, more importantly, empirical investigations can indicate the advantage of using predictions derived from non-linear models, over those from linear models.

Among other types of non-linear time series models, there

are models to represent the changes of variance along time (heteroskedasticity). These models are called autoregressive conditional heteroskedasticity (ARCH) and the collection comprises a wide variety of representation (GARCH, TARCH, EGARCH, FIGARCH, CGARCH, etc). Here changes in variability are related to, or predicted by, recent past values of the observed series. This is in contrast to other possible representations of locally-varying variability, where the variability might be modelled as being driven by a separate time-varying process, as in a doubly stochastic model.

In recent work on model-free analyses, wavelet transform based methods (for example locally stationary wavelets and wavelet decomposed neural networks) have gained favour. Multiscale (often referred to as multiresolution) techniques decompose a given time series, attempting to emphasize time dependence at multiple scales.

Notation

A number of different notations are in use for time-series analysis:

$$X = \{X_1, X_2 ...\}$$

is a common notation which specifies a time series X which is indexed by the natural numbers. Another common notation is:

$$Y = \{Y_t : t \in T\}.$$

Conditions

There are two sets of conditions under which much of the theory is built:

- Stationary process
- Ergodicity

However, ideas of stationarity must be expanded to consider two important ideas: strict stationarity and second-order stationarity. Both models and applications can be developed under each of these conditions, although the models in the latter case might be considered as only partly specified.

In addition, time-series analysis can be applied where the series are seasonally stationary and non-stationary.

Models

The general representation of an autoregressive model, well-known as AR(p), is

$$Y_t = \alpha_0 + \alpha_1 Y_{t-1} + \alpha_2 Y_{t-2} + \cdots + \alpha_p Y_{t-p} + \varepsilon_t$$

where the term ε_t is the source of randomness and is called white noise. It is assumed to have the following characteristics:

- $E[\varepsilon_t] = 0$
- $E[\varepsilon_t^2] = \sigma^2$
- $E[\varepsilon_t \varepsilon_s] = 0 \forall t \neq s$

With these assumptions, the process is specified up to second-order moments and, subject to conditions on the coefficients, may be second-order stationary.

If the noise also has a normal distribution, it is called normal white noise (denoted here by Normal-WN):

$$\{\varepsilon_t\}(t \in T) : \text{Normal} - \text{WN}$$

In this case the AR process may be strictly stationary, again subject to conditions on the coefficients.

LEAST SQUARES FITTING

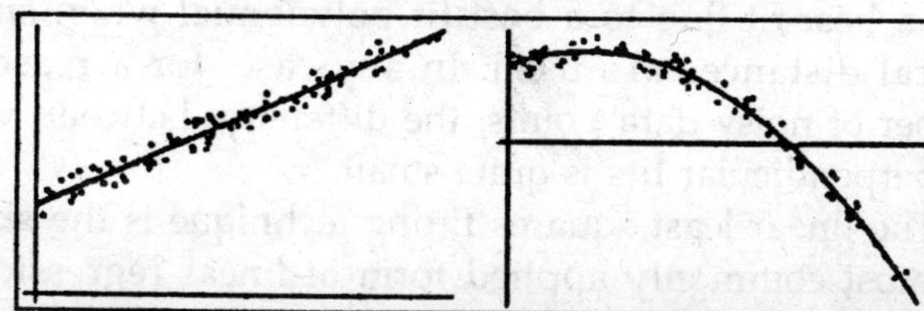

A mathematical procedure for finding the best-fitting curve to a given set of points by minimizing the sum of the squares of the offsets ("the residuals") of the points from the curve.

The sum of the squares of the offsets is used instead of the offset absolute values because this allows the residuals to be treated as a continuous differentiable quantity.

However, because squares of the offsets are used, outlying points can have a disproportionate effect on the fit, a property which may or may not be desirable depending on the problem at hand.

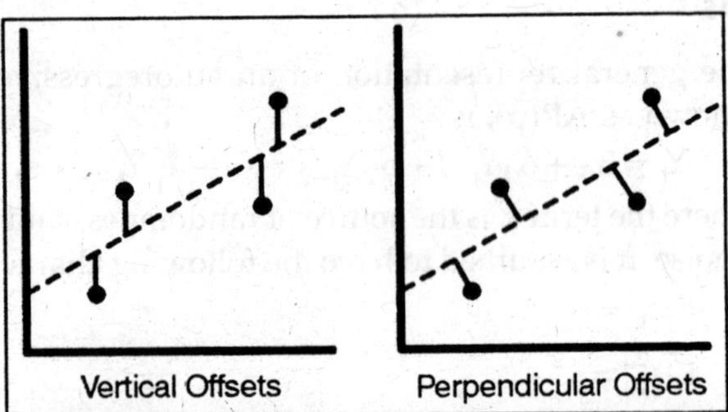

In practice, the vertical offsets from a line (polynomial, surface, hyperplane, etc.) are almost always minimized instead of the perpendicular offsets.

This provides a fitting function for the independent variable X that estimates y for a given x (most often what an experimenter wants), allows uncertainties of the data points along the x- and y-axes to be incorporated simply, and also provides a much simpler analytic form for the fitting parametres than would be obtained using a fit based on perpendicular offsets.

In addition, the fitting technique can be easily generalized from a best-fit line to a best-fit polynomial when sums of vertical distances are used. In any case, for a reasonable number of noisy data points, the difference between vertical and perpendicular fits is quite small.

The linear least squares fitting technique is the simplest and most commonly applied form of linear regression and provides a solution to the problem of finding the best fitting straight line through a set of points.

In fact, if the functional relationship between the two quantities being graphed is known to within additive or multiplicative constants, it is common practice to transform the data in such a way that the resulting line is a straight line, say by plotting T vs. $\sqrt{\ell}$ instead of T vs. ℓ in the case of analyzing the period Tof a pendulum as a function of its length ℓ. For this reason, standard forms for exponential, logarithmic, and power laws are often explicitly computed. The formulas for linear least squares fitting were independently derived by Gauss and Legendre.

For nonlinear least squares fitting to a number of unknown parametres, linear least squares fitting may be applied iteratively to a linearized form of the function until convergence is achieved. However, it is often also possible to linearize a nonlinear function at the outset and still use linear methods for determining fit parametres without resorting to iterative procedures.

This approach does commonly violate the implicit assumption that the distribution of errors is normal, but often still gives acceptable results using normal equations, a pseudoinverse, etc.

Depending on the type of fit and initial parametres chosen, the nonlinear fit may have good or poor convergence properties. If uncertainties (in the most general case, error ellipses) are given for the points, points can be weighted differently in order to give the high-quality points more weight.

Vertical least squares fitting proceeds by finding the sum of the squares of the vertical deviations R^2 of a set of n data points,

$$R^2 \equiv \sum [y_i - f(x_i, a_1, a_2, \ldots, a_n)]^2$$

from a function f. Note that this procedure does not minimize the actual deviations from the line (which would be measured perpendicular to the given function). In addition, although the unsquared sum of distances might seem a more appropriate quantity to minimize, use of the absolute value results in discontinuous derivatives which cannot be treated analytically.

The square deviations from each point are therefore summed, and the resulting residual is then minimized to find the best fit line. This procedure results in outlying points being given disproportionately large weighting.

The condition for R^2 to be a minimum is that,

$$\frac{\partial (R^2)}{\partial a_i} = 0$$

for i=..., n. For a linear fit,

$$f(a, b) = a + bx$$

so,

$$R^2(a,b) \equiv \sum_{i=1}^{n}[y_i - (a + bx_i)]^2$$

$$\frac{\partial(R^2)}{\partial a} = -2\sum_{i=1}^{n}[y_i - (a + b_{xi})] = 1$$

$$\frac{\partial(R^2)}{\partial b} = -2\sum_{i=1}^{n}[y_i - (a + bx_i)]x_i = 0$$

These lead to the equations,

$$na + b\sum_{i=1}^{n} x_i = \sum_{i=1}^{n} yi$$

$$a\sum_{i=1}^{n} x_i + b\sum_{i=1}^{n} x_i^2$$

In matrix form,

$$\begin{bmatrix} n & \sum_{i=1}^{n} x_i \\ \sum_{i=1}^{n} x_i & \sum_{i=1}^{n} x_i^2 \end{bmatrix}\begin{bmatrix} a \\ b \end{bmatrix} = \begin{bmatrix} \sum_{i=1}^{n} y_i \\ \sum_{i=1}^{n} x_i y_i \end{bmatrix}$$

so,

$$\begin{bmatrix} a \\ b \end{bmatrix} = \begin{bmatrix} n & \sum_{i=1}^{n} x_i \\ \sum_{i=1}^{n} x_i & \sum_{i=1}^{n} x_i^2 \end{bmatrix}^{-1}\begin{bmatrix} \sum_{i=1}^{n} y_i \\ \sum_{i=1}^{n} x_i y_i \end{bmatrix}.$$

The 2 × 2 matrix inverse is,

$$\begin{bmatrix} a \\ b \end{bmatrix} = \frac{1}{n\sum_{i=1}^{n} x_i^2 - \left(\sum_{i=1}^{n} x_i\right)^2}$$

$$\begin{bmatrix} \sum_{i=1}^{n} y_i \sum_{i=1}^{n} x_i^2 - \sum_{i=1}^{n} x_i \sum_{i=1}^{n} x_i y_i \\ n\sum_{i=1}^{n} x_i y_i - \sum_{i=1}^{n} x_i \sum_{i=1}^{n} y_i \end{bmatrix}$$

so

$$a = \frac{\sum_{i=1}^{n} y_i \sum_{i=1}^{n} x_i^2 - \sum_{i=1}^{n} x_i \sum_{i=1}^{n} x_i y_i}{n\sum_{i=1}^{n} x_i^2 - \left(\sum_{i=1}^{n} x_i\right)^2}$$

$$= \frac{\bar{y}\left(\sum_{i=1}^{n} x_i^2\right) - \bar{x}\sum_{i=1}^{n} x_i y_i}{\sum_{i=1}^{n} x_i^2 - n\bar{x}^2}$$

$$b = \frac{\sum_{i=1}^{n} x_i y_i - \sum_{i=1}^{n} x_i - \sum_{i=1}^{n} y_i}{n\sum_{i=1}^{n} x_i^2 - \left(\sum_{i=1}^{n} x_i\right)^2}$$

$$= \frac{\left(\sum_{i=1}^{n} x_i y_i\right) - n\bar{x}\,\bar{y}}{\sum_{i=1}^{n} x_i^2 - n\bar{x}^2}$$

These can be rewritten in a simpler form by defining the sums of squares,

$$SS_{xx} = \sum_{i=1}^{n} (x_i - \bar{x})^2$$

$$= \left(\sum_{i=1}^{n} x_i^2\right) - n\bar{x}^2$$

$$SS_{yy} = \sum_{i=1}^{n} (y_i - \bar{y})^2$$

$$= \left(\sum_{i=1}^{n} y_i^2\right) - n\bar{y}^2$$

$$SS_{xy} = \sum_{i=1}^{n} (x_i - \bar{x})(y_i - \bar{y})$$

$$= \left(\sum_{i=1}^{n} x_i y_i\right) - n\bar{x}\,\bar{y}$$

which are also written as,

$$\sigma_x^2 = \frac{SS_{xx}}{n}$$

$$\sigma_y^2 = \frac{SS_{yy}}{n}$$

$$\text{cov}(x, y) = \frac{SS_{xy}}{n}$$

Here, cov (x, y) is the covariance and σ_x^2 and σ_y^2 are variances. Note that the quantities,

$\sum_{i=1}^{n} x_i y_i$ and $\sum_{i=1}^{n} x_i^2$

can also be interpreted as the dot products,

$$\sum_{i=1}^{n} x_i^2 = x \cdot x$$

$$\sum_{i=1}^{n} x_i y_i = x \cdot y.$$

In terms of the sums of squares, the regression coefficient b is given by,

$$b = \frac{\text{cov}(x, y)}{\sigma_x^2} = \frac{SS_{xy}}{SS_{xx}}$$

and a is given in terms of b using () as,

$$a = \bar{y} - b\bar{x}$$

The overall quality of the fit is then parametreized in terms of a quantity known as the correlation coefficient, defined by,

$$r^2 = \frac{SS_{xy}^2}{SS_{xx} SS_{yy}}$$

which gives the proportion of SS_{yy} which is accounted for by the regression. Let $\hat{y}_i$ be the vertical coordinate of the best-fit line with x-coordinate x_i, so,

$$\hat{y}_i \equiv a + bx_i$$

then the error between the actual vertical point y_i and the fitted point is given by,

$$e_i \equiv y_i - \hat{y}_i$$

Now define s^2 as an estimator for the variance in e_i,

$$s^2 = \sum_{i=1}^{n} \frac{e_i^2}{n-2}$$

Then s can be given by,

$$S = \sqrt{\frac{SS_{yy} - bSS_{xy}}{n-2}} = \sqrt{\frac{SS_{yy} - \frac{SS_{xy}^2}{SS_{xx}}}{n-2}}$$

The standard errors for a and b are,

$$SE(a) = s\sqrt{\frac{1}{n} + \frac{\overline{x}^2}{SS_{xx}}}$$

$$SE(a) = \frac{s}{\sqrt{SS_{xx}}}$$

SEASONAL VARIATION

Seasonal variation is a component of a time series which is defined as the repetitive and predictable movement around the trend line in one year or less. It is detected by measuring the quantity of interest for small time intervals, such as days, weeks, months or quarters.

Organizations facing seasonal variations, like the motor vehicle industry, are often interested in knowing their performance relative to the normal seasonal variation. The same applies to the ministry of employment which expects unemployment to increase in June because recent graduates are just arriving into the job market and schools have also been given a vacation for the summer. The moot point is whether the increase is more or less than expected.

Organizations affected by seasonal variation need to identify and measure this seasonality to help with planning for temporary increases or decreases in Labour requirements, inventory, training, periodic maintenance, and so forth. Apart from these considerations, the organizations need to know if the variations they have experienced has been more or less would be expected given the usual seasonal variations.

REASONS FOR STUDYING SEASONAL VARIATION

There are several main reasons for studying seasonal variation:

- The description of the seasonal effect provides a better understanding of the impact this component has upon a particular series.
- After establishing the seasonal pattern, methods can be implemented to eliminate it from the time-series to study the effect of other components such as

cyclical and irregular variations. This elimination of the seasonal effect is referred to as deseasonalizing or seasonal adjustment of data.

- To project the past patterns into the future knowledge of the seasonal variations is a must for the prediction of the future trends.

ASSUMPTIONS

A decision maker or analyst can make one of the following assumptions when treating the seasonal component:

- The impact of the seasonal component is constant from year to year.
- The seasonal effect is changing slightly from year to year.
- The impact of the seasonal influence is changing dramatically.

SEASONAL INDEX

Seasonal variation is measured in terms of an index, called a seasonal index. It is an average that can be used to compare an actual observation relative to what it would be if there were no seasonal variation.

An index value is attached to each period of the time series within a year. This implies that if monthly data are considered there are 12 separate seasonal indices, one for each month. There can also be a further 4 index values for quarterly data. The following methods use seasonal indices to measure seasonal variations of a time-series data.

- Method of simple averages
- Ratio to trend method
- Ratio-to-moving average method
- Link relatives method

AN EXAMPLE

Now let us try to understand the measurement of seasonal variation by using the Ratio-to-Moving Average method. This technique provides an index to measure the degree of the Seasonal Variation in a time series.

The index is based on a mean of 100, with the degree of seasonality measured by variations away from the base. For example if we observe the hotel rentals in a winter resort, we find that the winter quarter index is 124.

The value 124 indicates that 124 per cent of the average quarterly rental occur in winter. If the hotel management records 1436 rentals for the whole of last year, then the average quarterly rental would be 359 (1436/4).

As the winter-quarter index is 124, we estimate the no. of winter rentals as follows:

359*(124/100)=445;

Here, 359 is the average quarterly rental. 124 is the winter-quarter index. 445 the seasonalized spring-quarter rental.

This is method is also called the percentage moving average method. In this method, the original data values in the time-series are expressed as percentages of moving averages. The steps and the tabulations are.

Steps

- Find the centred 12 monthly (or 4 quarterly) moving averages of the original data values in the time-series.
- Express each original data value of the time-series as a percentage of the corresponding centred moving average values obtained in step(1).In other words, in a multiplicative time-series model, we get(Original data values)/(Trend values) *100 = (T*C*S*I)/(T*C)*100 = (S*I) *100. This implies that the ratio-to-moving average represents the seasonal and irregular components.
- Arrange these percentages according to months or quarter of given years. Find the averages over all months or quarters of the given years.
- If the sum of these indices is not 1200, multiply then by a correction factor = 1200/ (sum of monthly indices). Otherwise, the 12 monthly averages will be considered as seasonal indices.

Let us calculate the seasonal index by the ratio-to-moving average method from the following data:

Year/Quarters	I	II	III	IV
1996	75	60	54	59
1997	86	65	63	80
1998	90	72	66	85
1999	100	78	72	93

Now calculations for 4 quarterly moving averages and ratio-to-moving averages are shown in the table.

Year	Quarter	Original Values(Y)	4 Figures Moving Total	4 Figures Moving Average	2 Figures Moving Total	2 Figures Moving Average(T)	Ratio-to-Moving Average(%)(Y)/(T)*100
1996	1	75					
–							
	2	60					
			248	62.00			
	3	54			126.75	63.375	85.21
			259	64.75			
	4	59			130.75	65.375	90.25
			264	66.00			
					134.25	67.125	128.12
1997	1	86					
			273	68.25			
	2	65			141.75	70.875	91.71
			294	73.50			
	3	63		148	74.00	84.13	
			298	74.50			
	4	80			150.75	75.375	106.14
			305	76.25			
1998	1	90			153.25	76.625	117.43
			308	77.00			
	2	72			155.25	77.625	92.75
			313	78.25			
	3	66			159.00	79.50	83.02
			323	80.75			
	4	85			163	81.50	104.29
			329	82.25			
1999	1	100			169.5	84.75	92.03
			335	83.75			
	2	78			169.50	84.75	92.03
			343	85.75			
	3	72					
	4	93					

Table. Calculation of seasonal index

Years/Quarters	1	2	3	4
1996	–	–	85.21	90.25
1997	128.12	91.71	85.13	106.14
1998	117.45	92.75	85.13	104.29
1999	117.99	92.03	–	–
Total	366.05	276.49	255.47	300.68

Seasonal Average	91.51	69.13	63.87	75.17
Adjusted Seasonal Average	122.07	92.22	85.20	100.30

Now the total of seasonal averages is 299.66. Therefore the corresponding correction factor would be 400/299.68 = 1.334.Each seasonal average is multiplied by the correction factor 1.334 to get the adjusted seasonal indices as shown in the table.

5

Theory

PROBABILITY THEORY

REASONING UNDER UNCERTAINTY

In many settings, we must try to understand what is going on in a system when we have imperfect or incomplete information.

- Two reasons why we might reason under uncertainty:
  - laziness (modeling every detail of a complex system is costly)
  - ignorance (we may not completely understand the system)
- *Example*: deploy a network of smoke sensors to detect _res in a building.

Our model will reect both laziness and ignorance:

  - We are too lazy to model what, besides _re, can trigger the sensors;
  - We are too ignorant to model how _re creates smoke, what density of smoke is required to trigger the sensors, etc.

USING PROBABILITY THEORY TO REASON UNDER UNCERTAINTY

- Probabilities quantify uncertainty regarding the occurrence of events.
- Are there alternatives? Yes, e.g., Dempster-Shafer Theory, disjunctive uncertainty, etc. (Fuzzy Logic is about imprecision, not uncertainty.)

- Why is Probability Theory better? de Finetti: Because if you do not reason just as to Probability Theory, you can be made to performance irrationally.
- Probability Theory is key to the study of action and communication:
  - Decision Theory combines Probability Theory with Utility Theory.
  - Information Theory is \the logarithm of Probability Theory".
- Probability Theory gives rise to many interesting and important philosophical questions (which as suggested, not cover).

THE ONLY PREREQUISITE: SET THEORY

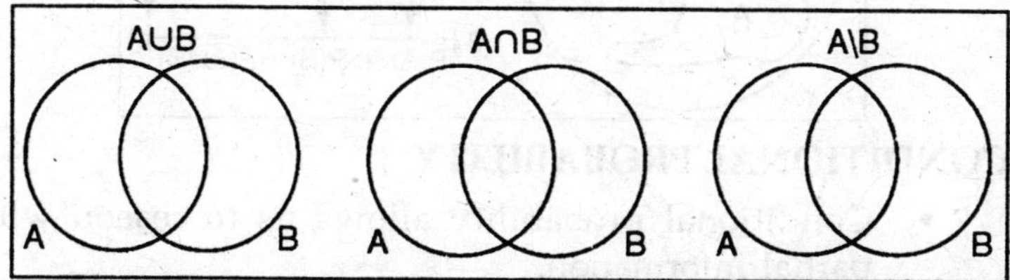

For simplicity, as suggested, work (mostly) with finite sets. The extension to countably infinite sets is not difficult. The extension to uncountably infinite sets requires Measure Theory.

PROBABILITY SPACES

- A probability space represents our uncertainty regarding an experiment.
- It has two parts:
  - The sample space, which is a set of outcomes;
  - The probability measure P, which is a real function of the subsets of.

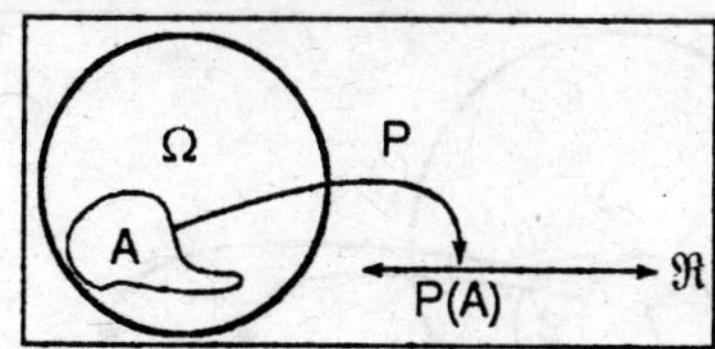

A set of outcomes A is called an event. P(A) represents

how likely it is that the experiment's actual outcome will be a member of A.

THE THREE AXIOMS OF PROBABILITY THEORY

- P(A) 0 for all events A
- P(Ω) = 1
- P(A B) = P(A) + P(B) for disjoint events A and B

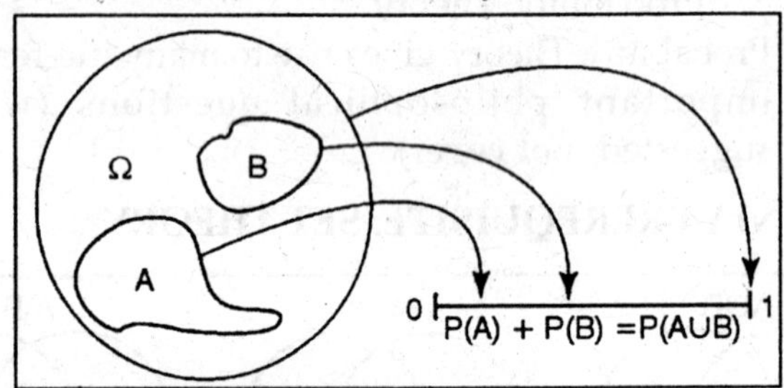

CONDITIONAL PROBABILITY

- Conditional probability allows us to reason with partial information.
- When P(B) > 0, the conditional probability of A given B is defined as

$$P(A \mid B) \triangleq \frac{P(A \cap B)}{P(B)}$$

This is the probability that A occurs, given we have observed B, i.e., that we know the experiment's actual outcome will be in B. It is the fraction of probability mass in B that also belongs to A.

- P(A) is called the a priori (or prior) probability of A and P(A B) is called the a posteriori probability of A given B.

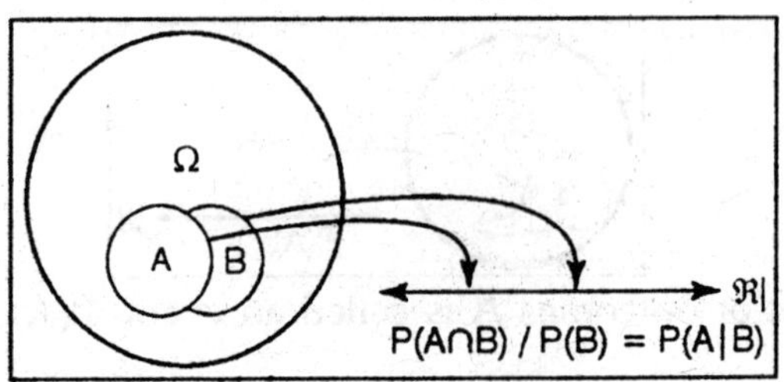

THE PRODUCT RULE

Start with the definition of conditional probability and multiply by P(A):

$$P(A \cap B) = P(A)P(B \mid A)$$

The probability that A and B both happen is the probability that A happens times the probability that B happens, given A has occurred.

THE CHAIN RULE

Apply the product rule repeatedly:

$$P\left(\cap_{i=1}^{k} A_i\right) = P(A_1)P(A_2 \mid A_1 \cap A_2) \cdots P\left(A_k \middle| \cap_{i=1}^{k-1} A_i\right)$$

The chain rule will become important later when we discuss conditional independence in Bayesian networks.

BAYES' RULE

Use the product rule both ways with P(A \ B) and divide by P(B):

$$P(A|B) = \frac{P(A|B)P(A)}{P(B)}$$

Bayes' rule translates causal knowledge into diagnostic knowledge. For example, if A is the event that a patient has a disease, and B is the event that she displays a symptom, then P(B A) describes a causal relationship, and P(A/B) describes a diagnostic one (that is usually hard to assess). If P(B/A), P(A) and P(B) can be assessed easily, then we get P(A/B) for free.

TREATMENT

Most introductions to probability theory treat discrete probability distributions and continuous probability distributions separately. The more mathematically advanced measure theory based treatment of probability covers both the discrete, the continuous, any mix of these two and more.

Discrete Probability Distributions

Discrete probability theory deals with events that occur in countable sample spaces.

Examples: Throwing dice, experiments with decks of cards,

and random walk. Classical definition: Initially the probability of an event to occur was defined as number of cases favourable for the event, over the number of total outcomes possible in an equiprobable sample space.

For example, if the event is"occurrence of an even number when a die is rolled", the probability is given by,

$$\frac{3}{6}=\frac{1}{2},$$

since 3 faces out of the 6 have even numbers and each face has the same probability of appearing.

Modern definition: The modern definition starts with a set called the sample space, which relates to the set of all possible outcomes in classical sense, denoted by

$$\Omega=\{x_1,x_2,...\}.$$

It is then assumed that for each element $x\in\Omega$, an intrinsic"probability" value f (x) is attached, which satisfies the following properties:

1. $f(x)\in[0,1]$ for all $x\in\Omega$;
2. $\sum_{x\in\Omega} f(x)=1$

That is, the probability function f(x) lies between zero and one for every value of x in the sample space Ω, and the sum of f(x) over all values x in the sample space Ω is equal to 1. An event is defined as any subset E of the sample space Ω. The probability of the event E is defined as,

$$P(E)=\sum_{x\in\Omega} f(x).$$

So, the probability of the entire sample space is 1, and the probability of the null event is 0. The function f(x) mapping a point in the sample space to the"probability" value is called a probability mass function abbreviated as pmf. The modern definition does not try to answer how probability mass functions are obtained; instead it builds a theory that assumes their existence.

Continuous Probability Distributions

Continuous probability theory deals with events that occur in a continuous sample space.

Classical definition: The classical definition breaks down when confronted with the continuous case.

Modern definition: If the outcome space of a random variable X is the set of real numbers ($\mathbb{R}$) or a subset thereof, then a function called the cumulative distribution function (or cdf) F exists, defined by $F(x) = P(X \le x)$. That is, F(x) returns the probability that X will be less than or equal to x.

The cdf necessarily satisfies the following properties.

- F is a monotonically non-decreasing, right-continuous function;
- $\lim_{x \to -\infty} F(x) = 0$;
- $\lim_{x \to \infty} F(x) = 1$.

If F is absolutely continuous, i.e., its derivative exists and integrating the derivative gives us the cdf back again, then the random variable X is said to have a probability density function or pdf or simply density,

$$f(x) = \frac{dF(x)}{dx}$$

For a set $E \subseteq \mathbb{R}$, the probability of the random variable X being in E is,

$$P(X \in E) = \int_{x \in E} dF(x)$$

In case the probability density function exists, this can be written as,

$$P(X \in E) = \int_{x \in E} f(x)dx$$

Whereas the pdf exists only for continuous random variables, the cdf exists for all random variables (including discrete random variables) that take values in $\mathbb{R}$.

These concepts can be generalized for multidimensional cases on $\mathbb{R}^n$ and other continuous sample spaces.

Measure-theoretic Probability Theory

The raison d'être of the measure-theoretic treatment of probability is that it unifies the discrete and the continuous cases, and makes the difference a question of which measure is used. Furthermore, it covers distributions that are neither discrete nor continuous nor mixtures of the two. An example of such distributions could be a mix of discrete and continuous distributions, for example, a random variable which is 0 with probability 1/2, and takes a random value from a normal distribution with probability 1/2. It can still be studied to some extent by considering it to have a pdf of $(\delta[x]+\varphi(x))/2$, where $\delta[x]$ is the Dirac delta function.

Other distributions may not even be a mix, for example, the Cantor distribution has no positive probability for any single point, neither does it have a density. The modern approach to probability theory solves these problems using measure theory to define the probability space:

Given any set Ω, (also called sample space) and a σ-algebra $\mathcal{F}$ on it, a measure $P(\Omega)=1$ defined on $\mathcal{F}$ is called a probability measure if $P(\Omega) = 1$

If $\mathcal{F}$ is the Borel σ-algebra on the set of real numbers, then there is a unique probability measure on $\mathcal{F}$ for any cdf, and vice versa. The measure corresponding to a cdf is said to be induced by the cdf. This measure coincides with the pmf for discrete variables, and pdf for continuous variables, making the measure-theoretic approach free of fallacies.

The probability of a set E in the σ-algebra $\mathcal{F}$ is defined as:

$$P(E)=\int_{\omega\in E}\mu_F(d\omega)$$

where the integration is with respect to the measure μ induced by F.

Along with providing better understanding and unification of discrete and continuous probabilities, measure-theoretic treatment also allows us to work on probabilities outside $\mathbb{R}^n$, as in the theory of stochastic processes. For example to study Brownian motion, probability is defined on a space of functions.

PROBABILITY DISTRIBUTIONS

Certain random variables occur very often in probability theory because they well describe many natural or physical processes. Their distributions therefore have gained special importance in probability theory. Some fundamental discrete distributions are the discrete uniform, Bernoulli, binomial, negative binomial, Poisson and geometric distributions. Important continuous distributions include the continuous uniform, normal, exponential, gamma and beta distributions.

CONVERGENCE OF RANDOM VARIABLES

In probability theory, there are several notions of convergence for random variables. They are listed in the order of strength, i.e., any subsequent notion of convergence in the list implies convergence according to all of the preceding notions.

Weak convergence: A sequence of random variables $X_1, X_2, \ldots,$ converges weakly to the random variable X if their respective cumulative distribution functions $F_1, F_2, \ldots,$ converge to the cumulative distribution function F of X, wherever F is continuous. Weak convergence is also called convergence in distribution.

Most common short hand notation:

$$X_n \xrightarrow{D} X$$

Convergence in probability: The sequence of random variables X_1, X_2... is said to converge towards the random variable X in probability if,

$$\lim_{x \to \infty} P(X_n - X) = 1$$

for every ε > 0.

Most common short hand notation:

$$X_n \xrightarrow{a.s.} X.$$

Strong convergence: The sequence of random variables X_1, X_2... is said to converge towards the random variable X strongly if,

$$P\left(\lim_{x \to \infty} X_n = X\right) = 1.$$

Strong convergence is also known as almost sure convergence.

Most common short hand notation:

$$X_n \xrightarrow{a.s.} X.$$

As the names indicate, weak convergence is weaker than

strong convergence. In fact, strong convergence implies convergence in probability, and convergence in probability implies weak convergence. The reverse statements are not always true.

LAW OF LARGE NUMBERS

Common intuition suggests that if a fair coin is tossed many times, then roughly half of the time it will turn up heads, and the other half it will turn up tails. Furthermore, the more often the coin is tossed, the more likely it should be that the ratio of the number of heads to the number of tails will approach unity.

Modern probability provides a formal version of this intuitive idea, known as the law of large numbers. This law is remarkable because it is nowhere assumed in the foundations of probability theory, but instead emerges out of these foundations as a theorem. Since it links theoretically-derived probabilities to their actual frequency of occurrence in the real world, the law of large numbers is considered as a pillar in the history of statistical theory.

The law of large numbers (LLN) states that the sample average,

$$\overline{X}_n = \frac{1}{n}\sum X_n$$

of X_1, X_2..., (independent and identically distributed random variables with finite expectation μ) converges towards the theoretical expectation μ. It is in the different forms of convergence of random variables that separates the weak and the strong law of large numbers,

$$\text{Weak law}: \overline{X}_n \xrightarrow{P} \mu \text{ for } n \to \infty$$

$$\text{Strong law}: \overline{X}_n \xrightarrow{a.s.} \mu \text{ for} \to \infty$$

It follows from LLN that if an event of probability p is observed repeatedly during independent experiments, the ratio of the observed frequency of that event to the total number of repetitions converges towards p.

Putting this in terms of random variables and LLN we have $Y_1, Y_2, \ldots$ are independent Bernoulli random variables taking values 1 with probability p and 0 with probability 1-p. E(Yi) = p for all i and it follows from LLN that,

$$\frac{\Sigma Y_n}{n}$$

converges to p almost surely.

CENTRAL LIMIT THEOREM

"The central limit theorem (CLT) is one of the great results of mathematics." It explains the ubiquitous occurrence of the normal distribution in nature.

The theorem states that the average of many independent and identically distributed random variables with finite variance tends towards a normal distribution irrespective of the distribution followed by the original random variables. Formally, let X_1, X_2... be independent random variables with mean μ and variance $\sigma^2 > 0$ Then the sequence of random variables,

$$Z_n = \frac{\sum_{i=1}^{n}(X_i - \mu)}{\sigma\sqrt{n}}$$

converges in distribution to a standard normal random variable.

BAYES' THEOREM

Bayes' Theorem is a simple mathematical formula used for calculating conditional probabilities. It figures prominently in subjectivist or Bayesian approaches to epistemology, statistics, and inductive logic.

Subjectivists, who maintain that rational belief is governed by the laws of probability, lean heavily on conditional probabilities in their theories of evidence and their models of empirical learning.

Bayes' Theorem is central to these enterprises both because it simplifies the calculation of conditional probabilities and because it clarifies significant features of subjectivist position. Indeed, the Theorem's central insight - that a hypothesis is confirmed by any body of data that its truth renders probable - is the cornerstone of all subjectivist methodology.

SIMPLE STATEMENT OF THEOREM

Bayes gave a special case involving continuous prior and posterior probability distributions and discrete probability distributions of data, but in its simplest setting involving only discrete distributions, Bayes' theorem relates the conditional and marginal probabilities of events A and B, where B has a non-vanishing probability:

$$P(A|B) = \frac{P(A|B)P(A)}{P(B)}$$

Each term in Bayes' theorem has a conventional name:

- P(A) is the prior probability or marginal probability of A. It is"prior" in the sense that it does not take into account any information about B.
- P(A I B) is the conditional probability of A, given B. It is also called the posterior probability because it is derived from or depends upon the specified value of B.
- P(B I A) is the conditional probability of B given A. It is also called the likelihood.
- P(B) is the prior or marginal probability of B, and acts as a normalizing constant.

Bayes' theorem in this form gives a mathematical representation of how the conditional probability of event A given B is related to the converse conditional probability of B given A.

Likelihood Functions and Continuous Prior and Posterior Distributions

Suppose a continuous probability distribution with probability density function f_Θ is assigned to an uncertain quantity Θ. (In the conventional language of mathematical probability theory Θ would be a"random variable") The probability that the event B will be the outcome of an experiment depends on Θ; it is P(B I Θ). As a function of Θ this is the likelihood function:

$$L(\theta) = P(B|\Theta = \theta)$$

Then the posterior probability distribution of Θ, i.e. the

conditional probability distribution of Θ given the observed data B, has probability density function

$$f_{\Theta}(\theta|B) = \text{constant} \cdot f_{\Theta}(\theta) L(B|\theta),$$

where the"constant" is a normalizing constant so chosen as to make the integral of the function equal to 1, so that it is indeed a probability density function. This is the form of Bayes' theorem actually considered by Thomas Bayes.

In other words, Bayes' theorem says: To get the posterior probability distribution, multiply the prior probability distribution by the likelihood function and then normalize.

More generally still, the new data B may be the value of an observed continuously distributed random variable X. The probability that it has any particular value is therefore 0. In such a case, the likelihood function is the value of a probability density function of X given Θ, rather than a probability of B given Θ:

$$L(\theta) = f_X(x|\Theta = \theta)$$

SIMPLE EXAMPLE

Suppose there is a school with 60% boys and 40% girls as students. The female students wear trousers or skirts in equal numbers; the boys all wear trousers. An observer sees a (random) student from a distance; all the observer can see is that this student is wearing trousers. What is the probability this student is a girl? The correct answer can be computed using Bayes' theorem.

The event A is that the student observed is a girl, and the event B is that the student observed is wearing trousers.

To compute P(A | B), we first need to know:

- P(A), or the probability that the student is a girl regardless of any other information. Since the observers sees a random student, meaning that all students have the same probability of being observed, and the fraction of girls among the students is 40%, this probability equals 0.4.
- P(B | A), or the probability of the student wearing trousers given that the student is a girl. As they are as likely to wear skirts as trousers, this is 0.5.

- P(B), or the probability of a (randomly selected) student wearing trousers regardless of any other information. Since half of the girls and all of the boys are wearing trousers, this is $0.5\times0.4 + 1\times0.6 = 0.8$.

Given all this information, the probability of the observer having spotted a girl given that the observed student is wearing trousers can be computed by substituting these values in the formula:

$$P(A|B)=\frac{P(A|B)P(A)}{P(B)}=\frac{0.5\times 0.4}{0.8}=0.25$$

Another, essentially equivalent way of obtaining the same result is as follows. Assume, for concreteness, that there are 100 students, 60 boys and 40 girls. Among these, 60 boys and 20 girls wear trousers. All together there are 80 trouser-wearers, of which 20 are girls. Therefore the chance that a random trouser-wearer is a girl equals 20/80 = 0.25. Put in terms of Bayes′ theorem, the probability of a student being a girl is 40/100, the probability that any given girl will wear trousers is 1/2. The product of these two is 20/100, but we know the student is wearing trousers, so you remove the 20 non trouser wearing students and receive a probability of (20/100)/(80/100), or 20/80.

It is often helpful when calculating conditional probabilities to create a simple table containing the number of occurrences of each outcome, or the relative frequencies of each outcome, for each of the independent variables. The table emphasizes the use of this method for the girl-or-boy example

	Girls	**Boys**	**Total**
Trousers	20	60	80
Skirts	20	0	20
Total	40	60	100

APPLICATION OF THE THEOREM

As a formal theorem, Bayes' theorem is valid in all common interpretations of probability. However, frequentist and Bayesian interpretations disagree on how (and to what)

probabilities are assigned. In the Bayesian interpretation, probabilities are rationally coherent degrees of belief, or a degree of belief in a proposition given a body of well-specified information.

Bayes' theorem can then be understood as specifying how an ideally rational person responds to evidence. In the frequentist interpretation, probabilities are the frequencies of occurrence of random events as proportions of a whole. Though his name has become associated with subjective probability, Bayes himself interpreted the theorem in an objective sense.

The theorem was given extra prominence by a theorem by physicist R. T. Cox which showed that any system of inference fitting certain requirements could be mapped onto probability. Bayes' Theorem has since found a wide variety of applications in science and engineering.

DERIVATION FROM CONDITIONAL PROBABILITIES

To derive the theorem, we start from the definition of conditional probability. The probability of event A given event B is,

$$P(A|B) = \frac{P(A \cap B)}{P(B)}$$

Equivalently, the probability of event B given event A is,

$$P(B|A) = \frac{P(A \cap B)}{P(A)}$$

Rearranging and combining these two equations, we find,

$$P(A|B) = P(B) = P(A \cap B) = P(B|A)P(A)$$

This lemma is sometimes called the product rule for probabilities. Discarding the middle term and dividing both sides by P(B), provided that neither P(B) nor P(A) is 0, we obtain Bayes' theorem:

$$P(A|B) = \frac{P(B|A)P(A)}{P(B)}$$

The lemma is symmetric in A and B and dividing by P(A),

provided that it is non-zero, gives a statement of Bayes' theorem where the two symbols have changed places.

Alternative Forms

Bayes' theorem is often completed by noting that, the Law of total probability,

$$P(B)=P(A\cap B)+P(A^c\cap B)=P(B|A)P(A)+P(B|A^c)P(A^c),$$

where A^C is the complementary event of A (often called"not A").

This results in the analogous form:

$$P(A|B)=\frac{P(B|A)P(A)}{P(B|A)P(A)+P(B|A^c)^P(A^c)}.$$

More generally, the law states that given a partition, i.e. {Ai}, of the event space,

$$P(B)=\sum_i P(B\cap A_i)=\sum_i P(B|A_i)P(A_i).$$

Thus, for any Ai in the partition, Bayes' theorem states that,

$$P(A_i|B)=\frac{P(B|A_i)P(A_i)}{P(B)}=\frac{P(B|A_i)P(A_i)}{\sum_j P(B|A_j)P(A_j)}.$$

In terms of Odds and Likelihood Ratio

Bayes' theorem can also be written neatly in terms of a likelihood ratio Λ and odds O as,

$$O(A|B)=O(A)\cdot\Lambda(A|B)$$

where O(A | B) are the (posterior) odds of A given B,

$$O(A|B)=\frac{P(A|B)}{P(A^c|B)}$$

O(A) are the (prior) odds of A by itself,

$$O(A)=\frac{P(A)}{P(A^c)}$$

and Λ(A | B) is the likelihood ratio.

For Probability Densities

There is also a version of Bayes' theorem for continuous distributions. It is somewhat harder to derive, since probability densities are not probabilities, so Bayes' theorem has to be established by a limit process. Bayes originally used the proposition to find a continuous posterior distribution given discrete observations. Bayes' theorem for probability densities is formally similar to the theorem for probabilities:

$$f_x(x|Y=y)=\frac{f_{X,Y}(x,y)}{f_Y(y)}=\frac{f_Y(y|X=x)fx(x)}{f_Y(y)}=$$

$$\frac{f_Y(y|X=x)f_x(x)}{\int_{\infty}^{\infty} fY(y|X=\xi)f_x(\xi)d\xi}$$

There is an analogous statement of the law of total probability, which is used in the denominator:

$$f_Y(y)=\int_{-\infty}^{\infty} f_Y(y|X=x)fx(x)dx.$$

As in the discrete case, the terms have standard names.

$$f_{X,Y}(x,y)$$

is the joint density function of X and Y,

$$f_X(x|Y=y)$$

is the posterior probability density function of X given Y = y,

$$f_Y(y|X=x)=L(x|y)$$

is (as a function of x) the likelihood function of X given Y = y, and

$$f_X(x)$$

and

$$f_Y(y)$$

are the marginal probability density functions of X and Y respectively, where $fX(x)$ is the prior probability density function of X.

Extensions

Theorems analogous to Bayes' theorem cover more than two events.

For example:

$$P(A|B\cap C)=\frac{P(A)P(B|A)P(C|A\cap B)}{P(B)P(C|B)}$$

This can be derived in a few steps from Bayes' theorem and the definition of conditional probability:

$$P(A|B\cap C)=\frac{P(A\cap B\cap C)}{P(B\cap C)}=\frac{P(C|A\cap B)P(A\cap B)}{P(B)P(C|B)}$$

$$=\frac{P(A)P(B|A)P(C|A\cap B)}{P(B)P(C|B)}$$

Similarly,

$$P(A|B\cap C)=\frac{P(B|A\cap C)P(A|C)}{P(B|C)}$$

can be regarded as a conditional Bayes' Theorem and can be derived as follows:

$$P(A|B\cap C)=\frac{P(A\cap B\cap C)}{P(B\cap C)}=\frac{P(B|A\cap C)P(A|C)P(C)}{P(C)P(B|C)}$$

$$=\frac{P(B|A\cap C)P(A|C)}{P(B|C)}.$$

A general strategy is to work with a decomposition of the joint probability, and to marginalize (integrate or sum) over the variables that are not of interest. Depending on the form of the decomposition, it may be possible to prove that some integrals must be 1, and thus they fall out of the decomposition; exploiting this property can reduce the computations very substantially. A Bayesian network, for example, specifies a factorization of a joint distribution of several variables in which the conditional probability of any one variable given the remaining ones takes a particularly simple form.

THEORETICAL DISTRIBUTIONS

DISTRIBUTIONS

Empirical distributions: distributions of scores that come from observation. Theoretical distributions: based on mathematical formulas and logic. Used in statistics to determine probabilities. When empirical and theoretical distributions correspond, you can use the theoretical one to determine probabilities of an outcome, which will lead to inferential statistics. The main point to retain is that theoretical distributions represent the"best estimate" of how events would actually occur. As with all estimates, a theoretical distribution may produce predictions that vary from observation, but will produce good estimates.

Three types of theoretical distributions: rectangular, binomial, and normal:

- *Probability*: The empirical approach to finding probability involves observing actual events, some of which are successes, other, failures. The ratio of successes to total events is expressed as a fraction, which you convert into a decimal number between.00 and 1.00. Probabilities are often stated as"chances in a hundred".
- Binomial distribution: Distribution of the frequency of events that can have only two possible outcomes
- Rectangular distribution (or uniform distribution): distribution in which all possible scores have the same probability of occurrence

THE NORMAL DISTRIBUTION

The normal distribution is a bell-shaped, theoretical distribution that predicts the frequency of occurrence of chance events. The probability of an event or a group of events corresponds to the area of the theoretical distribution associated with the event or group of event. The distribution is asymptotic: its line continually approaches but never reaches a specified limit. The curve is symmetrical: half of the total area is to the left, half to the right.

The total area under the theoretical distribution is always 1. Because it is so, there is a correspondence between area and probability.

The z scores: any distribution of scores, raw or theoretical, can be converted into a distribution of z scores, which is a score expressed in standard deviation unit. It is a mathematical way to modify an individual raw score so that the result conveys the score's relationship to the mean and standard deviation. It can be used both as a descriptive and as inferential statistic. Converting a raw score to a z score gives you a number that indicates the score's relative position in the distribution. If two raw scores are converted to z scores, it will give you their position in the distribution as well as their positions relative to each other. Once you determined the z score and found its position on the normal distribution, you can also determine probabilities of the score by looking at the corresponding area under the curve.

Inferential statistics are based on the concept of making decisions using distributions of sample statistics. Such a distribution is called a sampling distribution. The reason to use samples is that populations are usually too large or impossible to test. If you set up a sampling distribution, you can judge the relative position of your sample statistic as compared to the population statistic. With this information, you can make a judgment called a hypothesis test.

You generate sampling distributions by:

- Defining a population.
- Generating all possible samples of a given size from the population.
- Calculating a statistic (e.g., mean) for each sample.
- Plotting the distribution of the sample statistics.

Remember that the central limit theorem states that when an infinite number of successive random samples are taken from a population, the sampling distribution of the means of those samples will become approximately normally distributed with mean μ and standard deviation $\sigma/\sqrt{N}$ as the sample size (N) becomes larger, irrespective of the shape of the population distribution.

BINOMIAL

In elementary algebra, a binomial is a polynomial with two terms-the sum of two monomials-often bound by parenthesis or brackets when operated upon. It is the simplest kind of polynomial.

OPERATIONS ON SIMPLE BINOMIALS

- The binomial $a^2 - b^2$ can be factored as the product of two other binomials:

$$a^2 - b^2 = (a + b)(a - b).$$

This is a special case of the more general formula: $a^{n+1} - b^{n+1} = (a-b)\sum_{k=0}^{n} a^k b^{n-k}$.

- The product of a pair of linear binomials $(ax + b)$ and $(cx + d)$ is:

$$(ax + b)(cx + d) = acx^2 + adx + bcx + bd.$$

- A binomial raised to the nth power, represented as $(a + b)n$

can be expanded by means of the binomial theorem or, equivalently, using Pascal's triangle. Taking a simple example, the perfect square binomial $(p + q)^2$ can be found by squaring the first term, adding twice the product of the first and second terms and finally adding the square of the second term, to give $p^2 + 2pq + q^2$.

- A simple but interesting application of the cited binomial formula is the"(m,n)-formula" for generating Pythagorean triples: for $m < n$, let $a = n^2 < m^2$, $b = 2mn$, $c = n^2 + m^2$, then $a^2 + b^2 = c^2$.

POISSON DISTRIBUTION

In probability theory and statistics, the Poisson distribution (pronounced) (or Poisson law of small numbers) is a discrete probability distribution that expresses the probability of a number of events occurring in a fixed period of time if these events occur with a known average rate and independently of the time since the last event. (The Poisson distribution can also be used for the number of events in other specified intervals such as distance, area or volume.)

The distribution was first introduced by Siméon-Denis Poisson (1781-1840) and published, together with his probability theory, in 1838 in his work Recherches sur la probabilité des jugements en matière criminelle et en matière civile ("Research on the Probability of Judgments in Criminal and Civil Matters"). The work focused on certain random variables N that count, among other things, the number of discrete occurrences (sometimes called"arrivals") that take place during a time-interval of given length.

If the expected number of occurrences in this interval is λ, then the probability that there are exactly n occurrences (n being a non-negative integer, n = 0, 1, 2...) is equal to,

$$f(n;\lambda)=\frac{\lambda^{n}e^{-\lambda}}{n!},$$

where,

- e is the base of the natural logarithm (e = 2.71828...)
- n is the number of occurrences of an event - the probability of which is given by the function
- n! is the factorial of n
- λ is a positive real number, equal to the expected number of occurrences that occur during the given interval. For instance, if the events occur on average 4 times per minute, and you are interested in probability for n times of events occurring in a 10 minute interval, you would use as your model a Poisson distribution with λ = 10 × 4 = 40.

As a function of n, this is the probability mass function. The Poisson distribution can be derived as a limiting case of the binomial distribution.

The Poisson distribution can be applied to systems with a large number of possible events, each of which is rare. A classic example is the nuclear decay of atoms.

The Poisson distribution is sometimes called a Poissonian, analogous to the term Gaussian for a Gauss or normal distribution.

POISSON NOISE AND CHARACTERIZING SMALL OCCURRENCES

The parametre λ is not only the mean number of occurrences $\langle k \rangle$, but also its variance,

$$\sigma_k^2=\left\langle k^2\right\rangle-\langle k\rangle^2.$$

Thus, the number of observed occurrences fluctuates

about its mean λ with a standard deviation $\sigma_k = \sqrt{\lambda}$. These fluctuations are denoted as Poisson noise or (particularly in electronics) as shot noise. The correlation of the mean and standard deviation in counting independent, discrete occurrences is useful scientifically. By monitoring how the fluctuations vary with the mean signal, one can estimate the contribution of a single occurrence, even if that contribution is too small to be detected directly. For example, the charge e on an electron can be estimated by correlating the magnitude of an electric current with its shot noise.

If N electrons pass a point in a given time t on the average, the mean current is I = eN/ t; since the current fluctuations should be of the order $\sigma_I = e\sqrt{N}/t$ (i.e. the standard deviation of the Poisson process), the charge e can be estimated from the ratio σ_I^2/I. An everyday example is the graininess that appears as photographs are enlarged; the graininess is due to Poisson fluctuations in the number of reduced silver grains, not to the individual grains themselves. By correlating the graininess with the degree of enlargement, one can estimate the contribution of an individual grain (which is otherwise too small to be seen unaided). Many other molecular applications of Poisson noise have been developed, e.g., estimating the number density of receptor molecules in a cell membrane.

$$\Pr(N_t = k) = f(k;\lambda t) = \frac{e^{-\lambda t}(\lambda t)^k}{k!}$$

RELATED DISTRIBUTIONS

- If $X_1 \sim \text{Pois}(\lambda_1)$ and $X_2 \sim \text{Pois}(\lambda_2)$ then the difference $Y = X_1 - X_2$ follows a Skellam distribution.
- If $X_1 \sim \text{Pois}(\lambda_1)$ and $X_2 \sim \text{Pois}(\lambda_2)$ are independent, and Y = X1 + X2, then the distribution of X1 conditional on Y = y is a binomial. Specifically,

  $$X_1 | \sim (Y = y) \sim \text{Binom}(y, \lambda_1/(\lambda_1/\lambda_2)).$$

  More generally, if X_1, X_2..., X_n are independent Poisson random variables with parametres λ_1, λ_2..., λ_n then,

$$X_i \left| \sum_{j=1}^{n} X_j \sim \text{Binom}\left(\sum_{j=1}^{n} X_j, \frac{\lambda_i}{\sum_{j=1}^{n} \lambda_j} \right) \right.$$

- The Poisson distribution can be derived as a limiting case to the binomial distribution as the number of trials goes to infinity and the expected number of successes remains fixed. Therefore it can be used as an approximation of the binomial distribution if n is sufficiently large and p is sufficiently small. There is a rule of thumb stating that the Poisson distribution is a good approximation of the binomial distribution if n is at least 20 and p is smaller than or equal to 0.05, and an excellent approximation if $n \geq 100$ and $np \leq 10$.
- For sufficiently large values of λ, (say $\lambda > 1000$), the normal distribution with mean λ and variance λ (standard deviation $\sqrt{\lambda}$), is an excellent approximation to the Poisson distribution. If λ is greater than about 10, then the normal distribution is a good approximation if an appropriate continuity correction is performed, i.e., $P(X \leq x)$, where (lower-case) x is a non-negative integer, is replaced by $P(X \leq x + 0.5)$.

$$F_{\text{Poisson}}(x;\lambda) \approx F_{\text{normal}}\left(x;\mu = \lambda, \sigma^2 = \lambda\right)$$

- Variance-stabilizing transformation: When a variable is Poisson distributed, its square root is approximately normally distributed with expected value of about $\sqrt{\lambda}$ and variance of about 1/4. Under this transformation, the convergence to normality is far faster than the untransformed variable. Other, slightly more complicated, variance stabilizing transformations are available, one of which is Anscombe transform.
- If the number of arrivals in a given time interval [0,t] follows the Poisson distribution, with mean = λt, then the lengths of the inter-arrival times follow the Exponential distribution, with mean $1/\lambda$.

OCCURRENCE

The Poisson distribution arises in connection with Poisson processes. It applies to various phenomena of discrete properties (that is, those that may happen 0, 1, 2, 3... times during a given period of time or in a given area) whenever the probability of the phenomenon happening is constant in time or space.

Examples of events that may be modelled as a Poisson distribution include:

- The number of soldiers killed by horse-kicks each year in each corps in the Prussian cavalry. This example was made famous by Ladislaus Josephovich Bortkiewicz (1868-1931).
- The number of phone calls at a call centre per minute.
- Under an assumption of homogeneity, the number of times a web server is accessed per minute.
- The number of mutations in a given stretch of DNA after a certain amount of radiation.
- The proportion of cells that will be infected at a given multiplicity of infection.

HOW DOES THIS DISTRIBUTION ARISE? - THE LAW OF RARE EVENTS

In several of the examples-for example, the number of mutations in a given sequence of DNA-the events being counted are actually the outcomes of discrete trials, and would more precisely be modelled using the binomial distribution, that is,

$$X \sim B(n, p).$$

In such cases n is very large and p is very small (and so the expectation np is of intermediate magnitude). Then the distribution may be approximated by the less cumbersome Poisson distribution,

$$X \sim \text{Pois}(np).$$

This is sometimes known as the law of rare events, since each of the n individual Bernoulli events rarely occurs. The name may be misleading because the total count of success events in a Poisson process need not be rare if the parametre np is not small. For example, the number of telephone calls to a busy switchboard in one hour follows a Poisson distribution with the events appearing frequent to the operator, but they are rare from the point of the average member of the population who is very unlikely to make a call to that switchboard in that hour.

Proof

As suggested, prove that, for fixed λ, if,

$$X_n \sim B(n, \lambda/n); Y \sim \text{Pois}(\lambda).$$

then for each fixed k,

$$\lim_{x \to \infty} P(X_n = k) = P(Y = k).$$

Any Binomial random variable with large n and small p set λ = np. Note that the expectation E(Xn) = λ is fixed with respect to n.

First from calculus,

$$\lim_{x \to \infty} \left(1 - \frac{\lambda}{n}\right)^n = e^{-\lambda}$$

then since p = λ/ n in this case, we have,

$$\lim_{x \to \infty} P(X_n = k) = \lim_{x \to \infty} \binom{n}{k} p^k (1-p)^{n-k}$$

$$= \lim_{x \to \infty} \frac{n!}{(n-k)!k!} \left(\frac{\lambda}{n}\right)^k \left(1 - \frac{\lambda}{n}\right)^{n-k}$$

$$= \lim_{x \to \infty} \underbrace{\left[\frac{n!}{n^k (n-k)!}\right]}_{An} \left(\frac{\lambda^k}{k!}\right) \underbrace{\left(1 - \frac{\lambda}{n}\right)^n}_{\to \exp(-\lambda)} \underbrace{\left(1 - \frac{\lambda}{n}\right)^n}_{\to 1}$$

$$= \left[\lim_{x \to \infty} An\right] \left(\frac{\lambda^k}{k!}\right) \exp(-\lambda)$$

Next, note that,

$$A_n = \frac{n!}{n^k (n-k)!}$$

$$= \frac{n \cdot (n-1) \cdots (n-(k-1))}{n^k}$$

$$= 1 \cdot \left(1 - \frac{1}{n}\right) \cdots \left(1 - \frac{k-1}{n}\right)$$

$$\to 1 \cdot 1 \cdots 1 = 1$$

where we have taken the limit of each of the terms independently, which is permitted since there is a fixed number of terms with respect to n (there are k of them). Consequently, we have shown that,

$$\lim_{x \to \infty} P(X_n = k) = \frac{\lambda^k \exp(-\lambda)}{k!} = P(Y = k).$$

Generalization

We have shown that if,

$$X_n \sim B(n, p_n); Y \sim \text{Pois}(\lambda)$$

where $p_n = \lambda / n$, then $X_n \to Y$ in distribution. This holds in the more general situation that pn is any sequence such that,

$$\lim_{x \to \infty} np_n = \lambda$$

2-dimensional Poisson Process

$$P(N(D) = k) = \frac{(\lambda|D|)^k e^{-\lambda|D|}}{k!}$$

where

- 1 (e = 2.71828...)
- k is the number of occurrences of an event - the probability of which is given by the function
- k! is the factorial of k
- D is the 2-dimensional region
- N(D) is the number of points in the process in region D

PROPERTIES

- The expected value of a Poisson-distributed random variable is equal to λ and so is its variance. The higher moments of the Poisson distribution are Touchard polynomials in λ, whose coefficients have a combinatorial meaning. In fact, when the expected value of the Poisson distribution is 1, then Dobinski's formula says that the nth moment equals the number of partitions of a set of size n.

- The mode of a Poisson-distributed random variable with non-integer λ is equal to $[\lambda]$, which is the largest integer less than or equal to λ. This is also written as floor(λ). When λ is a positive integer, the modes are λ and $\lambda - 1$.
- *Sums of Poisson-distributed random variables*: If $X_i \sim \text{Pois}(\lambda_i)$ follow a Poisson distribution with parametre λ_i and Xi are independent, then

$$Y = \sum_{i=1}^{N} X_i \sim \text{Pois}\left(\sum_{i=1}^{N} \lambda_i\right)$$

also follows a Poisson distribution whose parametre is the sum of the component parametres. A converse is Raikov's theorem, which says that if the sum of two independent random variables is Poisson-distributed, then so is each of those two independent random variables.

- The sum of normalised square deviations is approximately distributed as chi-square if the mean is of a moderate size ($\lambda > 5$ is suggested). If $X_1 \ldots, X_N$ are observations from independent Poisson distributions with means $\lambda_1 \ldots, \lambda_N$ then,

$$\sum_{i=1}^{N} \frac{(X_i - \lambda_i)^2}{\lambda_i} \sim X^2$$

- The moment-generating function of the Poisson distribution with expected value λ is

$$E\left(e^{tX}\right) = \sum_{k=0}^{\infty} e^{tk} f(k;\lambda) = \sum_{k=0}^{\infty} e^{tk} \frac{\lambda^k e^{-\lambda}}{k!} = e^{\lambda\left(e^t - 1\right)}$$

- All of the cumulants of the Poisson distribution are equal to the expected value λ. The nth factorial moment of the Poisson distribution is λ^n.
- The Poisson distributions are infinitely divisible probability distributions.

- The directed Kullback-Leibler divergence between Pois(λ) and Pois(λ_0) is given by,

$$D_{KL}(\lambda \| \lambda_0) = \lambda_0 - \lambda + \lambda \log \frac{\lambda}{\lambda_0}$$

PARAMETRE ESTIMATION

Maximum likelihood

Given a sample of n measured values k_i we wish to estimate the value of the parametre λ of the Poisson population from which the sample was drawn. To calculate the maximum likelihood value, we form the log-likelihood function

$$L(\lambda) = \text{In} \prod_{i=1}^{n} f(k_i | \lambda)$$

$$= \sum_{i-1}^{n} \text{In}\left(\frac{e^{-\lambda} \lambda^{ki}}{k_i !}\right)$$

$$= -n\lambda + \left(\sum_{i=1}^{n} k_i\right) \text{In}(\lambda) - \sum_{i=1}^{n} \text{In}(k_i !)$$

Take the derivative of L with respect to λ and equate it to zero:

$$\frac{d}{d\lambda} L(\lambda) = 0 \Leftrightarrow -n + \left(\sum_{i=1}^{n} k_i\right) \frac{1}{\lambda} = 0$$

Solving for λ yields a stationary point, which if the second derivative is negative is the maximum-likelihood estimate of λ:

$$\hat{\lambda}_{MLE} = \frac{1}{n} \sum_{i=1}^{n} k_i$$

Checking the second derivative, it is found that it is negative for all λ and ki greater than zero, therefore this stationary point is indeed a maximum of the initial likelihood function:

$$\frac{\partial^2 L}{\partial \lambda^2} = \sum_{i=1}^{n} -\lambda^{-2} k_i$$

Since each observation has expectation λ so does this sample mean. Therefore it is an unbiased estimator of λ. It is also an efficient estimator, i.e. its estimation variance achieves the Cramér-Rao lower bound (CRLB). Hence it is MVUE. Also it can be proved that the sample mean is complete and sufficient statistic for λ.

Bayesian Inference

In Bayesian inference, the conjugate prior for the rate parametre λ of the Poisson distribution is the Gamma distribution.

Let,

$$\lambda \sim \mathrm{Gamma}(\alpha, \beta)$$

denote that λ is distributed just as to the Gamma density g parametreized in terms of a shape parametre α and an inverse scale parametre β:

$$g(\lambda|\alpha,\beta) = \frac{\beta^{\alpha}}{\Gamma(\alpha)} \lambda^{\alpha-1} e^{-\beta\lambda} \text{ for } \lambda > 0$$

Then, given the same sample of n measured values k_i as before, and a prior of Gamma(α, β), the posterior distribution is,

$$\lambda \sim \mathrm{Gamma}\left(\alpha + \sum_{i=1}^{n} k_i, \beta + n\right).$$

The posterior mean E[λ] approaches the maximum likelihood estimate $\hat{\lambda}$ in the limit as $\alpha \to 0, \beta \to 0$.

The posterior predictive distribution of additional data is a Gamma-Poisson (i.e. negative binomial) distribution.

THE"LAW OF SMALL NUMBERS"

The word law is sometimes used as a synonym of probability distribution, and convergence in law means convergence in distribution.

The Poisson distribution is sometimes called the law of small numbers because it is the probability distribution of the number of occurrences of an event that happens rarely but has very many opportunities to happen.

Some historians of mathematics have argued that the Poisson distribution should have been called the Bortkiewicz distribution.

NORMAL DISTRIBUTION

In many natural processes, random variation conforms to a particular probability distribution known as the normal distribution, which is the most commonly observed probability distribution. Mathematicians de Moivre and Laplace used this distribution in the 1700's. In the early 1800's, German mathematician and physicist Karl Gauss used it to Analyse astronomical data, and it consequently became known as the Gaussian distribution among the scientific community.

The shape of the normal distribution resembles that of a bell, so it sometimes is referred to as the"bell curve", an example of which follows:

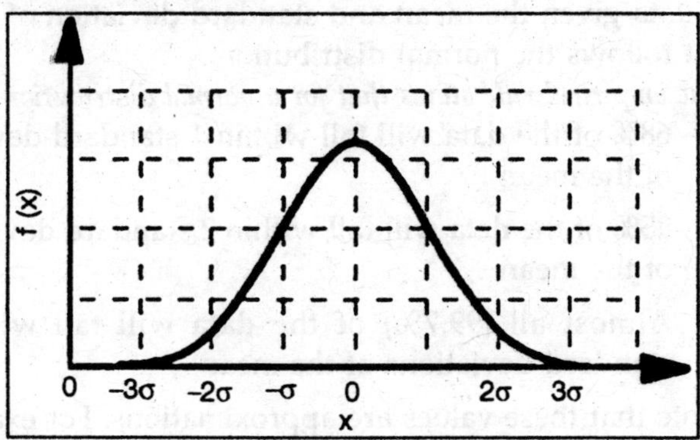

Fig. Normal Distribution

The curve is for a data set having a mean of zero. In general, the normal distribution curve is described by the following probability density function:

$$f(x) = \frac{1}{\sigma\sqrt{2\pi}} e^{-\frac{1}{2}\left(\frac{x-\mu}{\sigma}\right)^2}$$

BELL CURVE CHARACTERISTICS

The bells curve has the following characteristics:

- Symmetric
- Unimodal
- Extends to +/– infinity
- Area under the curve = 1

COMPLETELY DESCRIBED BY TWO PARAMETRES

The normal distribution can be completely specified by two parametres:

- Mean
- Standard deviation

If the mean and standard deviation are known, then one essentially knows as much as if one had access to every point in the data set.

THE EMPIRICAL RULE

The empirical rule is a handy quick estimate of the spread of the data given the mean and standard deviation of a data set that follows the normal distribution.

The empirical rule states that for a normal distribution:

- 68% of the data will fall within 1 standard deviation of the mean
- 95% of the data will fall within 2 standard deviations of the mean
- Almost all (99.7%) of the data will fall within 3 standard deviations of the mean

Note that these values are approximations. For example, the normal curve probability density function, 95% of the data will fall within 1.96 standard deviations of the mean; 2 standard deviations is a convenient approximation.

NORMAL DISTRIBUTION AND THE CENTRAL LIMIT THEOREM

The normal distribution is a widely observed distribution. Furthermore, it frequently can be applied to situations in which the data is distributed very differently.

This extended applicability is possible because of the central limit theorem, which states that regardless of the distribution of the population, the distribution of the means of random samples approaches a normal distribution for a large sample size.

APPLICATIONS TO BUSINESS ADMINISTRATION

The normal distribution has applications in many areas of business administration.

For example:

- Modern portfolio theory commonly assumes that the returns of a diversified asset portfolio follow a normal distribution.
- In operations management, process variations often are normally distributed.
- In human resource management, employee performance sometimes is considered to be normally distributed.

The normal distribution often is used to describe random variables, especially those having symmetrical, unimodal distributions. In many cases however, the normal distribution is only a rough approximation of the actual distribution. For example, the physical length of a component cannot be negative, but the normal distribution extends indefinitely in both the positive and negative directions. Nonetheless, the resulting errors may be negligible or within acceptable limits, allowing one to solve problems with sufficient accuracy by assuming a normal distribution.

SAMPLING DISTRIBUTION

In statistics, a sampling distribution or finite-sample distribution is the distribution of a given statistic based on a random sample of size n. It may be considered as the distribution of the statistic for all possible samples of a given size. The sampling distribution depends on the underlying distribution of the population, the statistic being considered, and the sample size used. The sampling distribution is frequently opposed to the asymptotic distribution, which corresponds to the limit case $n \to \infty$.

For example, consider a normal population with mean μ and variance σ^2. Assume we repeatedly take samples of a given size from this population and calculate the arithmetic mean $\overline{x}$ for each sample - this statistic is called the sample mean.

Each sample will have its own average value, and the distribution of these averages will be called the"sampling distribution of the sample mean". This distribution will be normal $N\left(\mu, \sigma^2 / n\right)$ since the underlying population is normal.

This was an example of a simple statistic taken from one of the simplest statistical populations. For other statistics and other populations the formulas are frequently more complicated, and oftentimes they don't even exist in closed-form. In such cases the sampling distributions may be approximated through Monte-Carlo simulations, bootstrap method, or asymptotic distribution theory.

The standard deviation of the sampling distribution of the statistic is referred to as the standard error of that quantity. For the case where the statistic is the sample mean, the standard error is:

$$\sigma_{\bar{x}} = \frac{\sigma}{\sqrt{n}}$$

where σ is the standard deviation of the population distribution of that quantity and n is the size (number of items) in the sample.

A very important implication of this formula is that you must quadruple the sample size (4×) to achieve half (1/2) the measurement error. When designing statistical studies where cost is a factor, this may have a factor in understanding cost-benefit tradeoffs.

Alternatively, consider the sample median from the same population. It has a different sampling distribution which is generally not normal (but may be close under certain circumstances).

STANDARD ERROR

The standard error of a statistic is the standard deviation of the sampling distribution of that statistic. Standard errors are important because they reflect how much sampling fluctuation a statistic will show. The inferential statistics involved in the construction of confidence intervals and significance testing are based on standard errors. The standard error of a statistic depends on the sample size. In general, the larger the sample size the smaller the standard error. The standard error of a statistic is usually designated by the Greek letter sigma (σ) with a subscript indicating the statistic. For instance, the standard error of the mean is indicated by the symbol: σ_M.

STANDARD ERROR OF THE MEAN

The standard error of the mean is designated as: σ_M. It is the standard deviation of the sampling distribution of the mean.

The formula for the standard error of the mean is:

$$\sigma_M = \frac{\sigma}{\sqrt{N}}$$

where σ is the standard deviation of the original distribution and N is the sample size (the number of scores each mean is based upon). This formula does not assume a normal distribution. However, many of the uses of the formula do assume a normal distribution. The formula shows that the larger the sample size, the smaller the standard error of the mean. More specifically, the size of the standard error of the mean is inversely proportional to the square root of the sample size. A graph of the effect of sample size on the standard error for a standard deviation of 10 is shown:

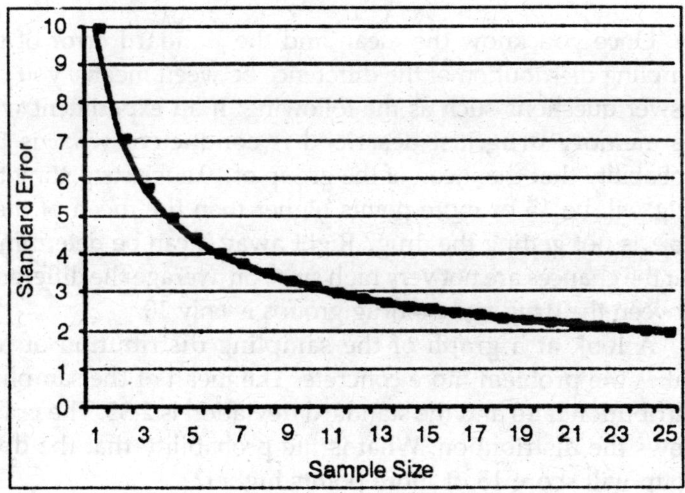

The function levels off. Increasing the sample size by a few subjects makes a big difference when the sample size is small but makes much less of a difference when the sample size is large. Notice that the graph is consistent with the formulas. If σ_M is 10 for a sample size of 1 then σ_M should be equal to,

$$\frac{10}{\sqrt{25}}$$

for a sample size of 25. When s is used as an estimate of σ, the estimated standard error of the mean is,

$$S_M \frac{S}{\sqrt{N}}.$$

The standard error of the mean is used in the computation of confidence intervals and significance tests for the mean.

SAMPLING DISTRIBUTION, DIFFERENCE BETWEEN INDEPENDENT MEANS

If $n_1 = 10$ and $n_2 = 8$ then,

$$\sigma^2_{M_d} = \frac{25}{10} + \frac{24}{8} = 5.5.$$

Finally the standard error of M_d is simply the square root of the variance of the sampling distribution of M_d. So,

$$\sigma_{M_d} = \sqrt{\frac{\sigma_1^2}{n_1} + \frac{\sigma_2^2}{n_2}} = 2.35$$

Once you know the mean and the standard error of the sampling distribution of the difference between means, you can answer questions such as the following: If an experiment with the memory drug just described is conducted, what is the probability that the mean of the group of 10 subjects getting the drug will be 15 or more points higher than the mean of the 8 subjects not getting the drug? Right away it can be determined that the chances are not very high since on average the difference between the drug and no-drug groups is only 10.

A look at a graph of the sampling distribution of Md makes the problem more concrete. The mean of the sampling distribution is 10 and the standard deviation is 2.35. The graph shows the distribution. What is the probability that the drug group will score 15 or more points higher?

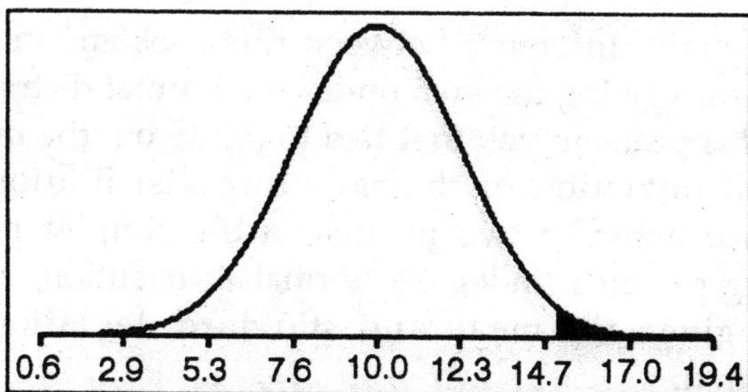

The blue region of the graph is 15 or higher and is a small portion of the area. The probability can be determined by computing the number of standard deviations the mean 15 is. Since the mean is 10 and the standard deviation is 2.35, 15 is (15-10)/2.35 = 2.13 standard deviations above the mean. From a z table it can be determined that 0.983 of the area is below 2.13; therefore, 0.017 of the area is above. It follows that the probability of a difference between the drug and no-drug means of 15 or larger is.017.

As shown, the z table calculator allows you to compute this value directly without using z scores.

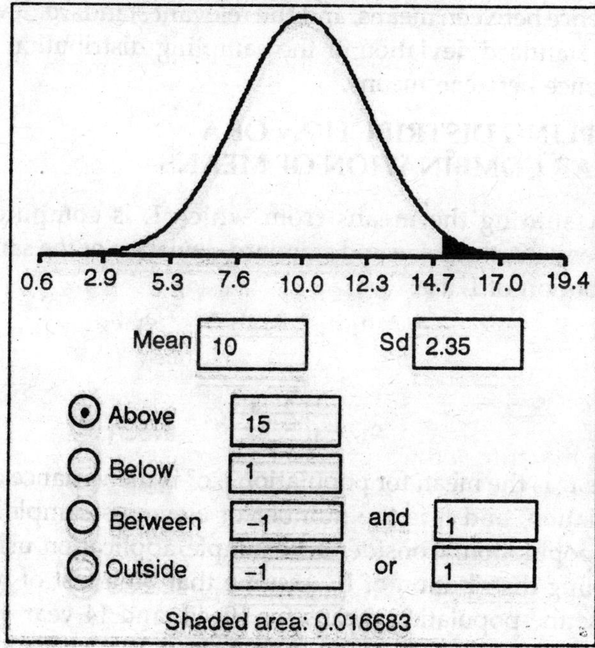

The main difference between this problem and simple problems involving the area under the normal distribution is that in this problem you first had to figure out the mean and standard deviation of the sampling distribution of the difference between two means. In the simpler problems involving the area under the normal distribution, you were always given the mean and standard deviation of the distribution.

The formula used there:

$$z = \frac{X - \mu}{\sigma}$$

is used here in a different form:

$$Z = \frac{M_d - \mu_{M_d}}{\sigma_{M_d}}$$

Since the problem now concerns differences between means, the relevant statistic is the difference between means obtained in the sample (experiment), the relevant population mean (μ) is the mean of the sampling distribution of the difference between means, and the relevant standard deviation is the standard deviation of the sampling distribution of the difference between means.

SAMPLING DISTRIBUTION OF A LINEAR COMBINATION OF MEANS

Assuming the means from which L is computed are independent, the mean and standard deviation of the sampling distribution of L are:

$$\mu_L = a_1 \mu_1 + a_2 \mu_2 + \ldots + a_k \mu_k$$

and

$$\sigma_L = \sqrt{\frac{\sum a_i^2}{n} \sigma^2}$$

where μ_i is the mean for population i, σ^2 is the variance of each population, and n is the number of elements sampled from each population. Consider an example application using the sampling distribution of L. Assume that on a test of reading ability, the population means for 10, 12, and 14 year olds are

60, 68, and 80 respectively. Further assume that the variance within each of these three populations is 100. Then, $\mu_1 = 60$, $\mu_2 = 68$, $\mu_3 = 80$, and $\sigma^2 = 100$. If eight 10-year-olds, eight 12- year-olds and eight 14-year-olds are sampled randomly, what is the probability that the mean for the 14 year olds will be 15 or more points higher than the average of the means for the two younger groups?

Or, symbolically, what is the probability that:

$$M_3 - \frac{M_1 + M_2}{2} \geq 15?$$

The answer lies in the sampling distribution of,

$$M_3 - \frac{M_1 + M_2}{2}.$$

Letting $a_1 = -.5$, $a_2 = -.5$, and $a_3 = 1$,

$$L = M_3 - \frac{M_1 + M_2}{2}.$$

The mean and standard deviation of the sampling distribution of L for this example are: $\mu_L = a_1\,\mu_1 + a_2\,\mu_2 + \ldots + a_k\,\mu_k = (-.5)(60) + (-.5)(68) + (1)(80) = 16$ and,

$$\sigma_L = \sqrt{\frac{\sum a_i^2}{n}\sigma^2} = \sqrt{\frac{(-0.5)^2 + (-0.5)^2 + 1^2}{8}\sigma^2} = 4.33$$

The sampling distribution of L therefore has a mean of 16 and a standard deviation of 4.33. The question is, what is the probability of getting a value of L greater than or equal to 15?

The formula:

$$z = \frac{L - \mu_L}{\sigma_L} = (15 - 16)/4.33 = -0.23$$

can be used to find out how many standard deviations above μ_L an L of 15 is. Using a z table, it can be determined that 0.41 of the time a z of –0.23 or lower would occur. Therefore the probability of a z of –0.23 or higher occurring is (1 – 0.41) = 0.59. So, the probability that,

$$M_3 = \frac{M_1 + M_2}{2}$$

is greater than 15 is 0.59.

SAMPLING DISTRIBUTION OF PEARSON'S R

Just like any other statistic, Pearson's r has a sampling distribution. If N pairs of scores were sampled over and over again the resulting Pearson r's would form a distribution. When the absolute value of the correlation in the population is low then the sampling distribution of Pearson's r is approximately normal. However, with high values of correlation, the distribution has a negative skew. The graph shows the sampling distribution of Pearson's r when the population correlation is 0.60 and when N = 12. The negative skew is apparent.

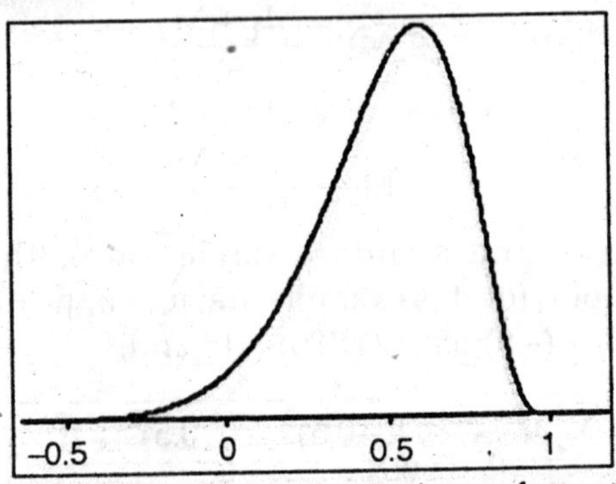

A transformation called Fisher's z' transformation converts Pearson's r to a value that is normally distributed and with a standard error of:

$$\sigma_z = \frac{1}{\sqrt{N-3}}$$

SAMPLING DISTRIBUTION OF MEDIAN

The standard error of the median for large samples and normal distributions is:

$$\sigma_{median} = 1.253\frac{\sigma}{\sqrt{N}}$$

Thus, the standard error of the median is about 25% larger than that for the mean. It is thus less efficient and more subject to sampling fluctuations. This formula is fairly accurate even for small samples but can be very wrong for extremely non-normal distributions. For non-normal distributions, the standard error of the median is difficult to compute. The sampling distribution simulation can be used to explore the sampling distribution of the median for non-normal distributions.

SAMPLING DISTRIBUTION OF THE STANDARD DEVIATION

The standard error of the standard deviation is approximately $\sigma_S = .71\sigma / \sqrt{N}$.

The approximation is more accurate for larger sample sizes ($N > 16$) and when the population is normally distributed. Try this simulation to explore the accuracy of the approximation.

The distribution of the standard deviation is positively skewed for small N but is approximately normal if N is 25 or greater. Thus, procedures for calculating the area under the normal curve work for the sampling distribution of the standard deviation as long as N is at least 25 and the distribution is approximately normal.

SAMPLING DISTRIBUTION OF A PROPORTION

Assume that 0.80 of all third grade students can pass a test of physical fitness. A random sample of 20 students is chosen: 13 passed and 7 failed.

The parametre π is used to designate the proportion of subjects in the population that pass (.80 in this case) and the statistic p is used to designate the proportion who pass in a sample (13/20 =.65 in this case).

The sample size (N) in this example is 20. If repeated samples of size N where taken from the population and the proportion passing (p) were determined for each sample, a distribution of values of p would be formed. If the sampling went on forever, the distribution would be the sampling distribution of a proportion.

The sampling distribution of a proportion is equal to the binomial distribution. The mean and standard deviation of the binomial distribution are:

$$\mu = \pi$$

and

$$\sigma_p = \sqrt{\frac{\pi(1-\pi)}{N}}.$$

For the present example, N = 20, π = 0.80, the mean of the sampling distribution of p (μ) is.8 and the standard error of p (σ_p) is 0.089. The shape of the binomial distribution depends on both N and π. With large values of N and values of π in the neighbourhood of.5, the sampling distribution is very close to a normal distribution.

The plot shown here is the sampling distribution for the present example. The distribution is not far from the normal distribution, although it does have some negative skew.

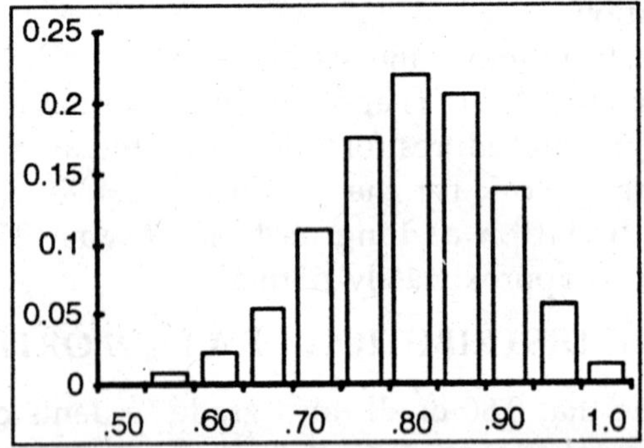

Assume that for the population of people applying for a job at a bank in a major city.40 are able to pass a basic literacy test required to get the job. Out of a group of 20 applicants, what is the probability that 50% or more of them will pass? This problem involves the sampling distribution of p with π =.40 and N = 20. The mean of the sampling distribution is π =.40.

The standard deviation is:

$$\sigma_p = \sqrt{\frac{\pi(1-\pi)}{N}} = \sqrt{\frac{0.40(1-0.40)}{20}} = 0.11$$

Using the normal approximation, a proportion of.50 is: (.50-.40)/.11 = 0.909 standard deviations above the mean. From a z table it can be calculated that 0.818 of the area is a z of 0.909. Therefore the probability that 50% or more will pass the literacy test is only about 1 – 0.818 = 0.182.

Correction for Continuity

Since the normal distribution is a continuous distribution, the probability that a sample value will exactly equal any specific value is zero. However, this is not true when the normal distribution is used to approximate the sampling distribution of a proportion. A correction called the"correction for continuity" can be used to improve the approximation.

The basic idea is that to estimate the probability of, 10 successes out of 20 when π is 0.4, one should compute the area between 9.5 and 10.5 as shown.

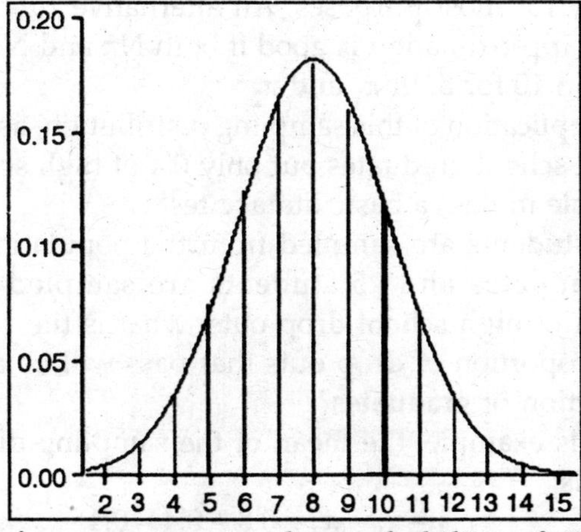

Therefore to compute the probability of 10 or more successes, compute the area above 9.5 successes. In terms of proportions, 9.5 successes is 9.5/20 = 0.475. Therefore, 9.5 = (0.475 - 0.40)/.11 = 0.682 standard deviations above the mean. The probability of being 0.682 or more standard deviations above the mean is 0.247 rather than the 0.182 that was obtained previously.The exact answer calculated using the binomial distribution is 0.245. For small sample sizes the correction can make a much bigger difference than it did here.

SAMPLING DISTRIBUTION OF THE DIFFERENCE BETWEEN TWO PROPORTIONS

The mean of the sampling distribution of the difference between two independent proportions ($p_1 - p_2$) is:

$$\mu_{P_1-P_2} = \pi_1 - \pi_2.$$

The standard error of $p_1 - p_2$ is:

$$\sigma_{P_1-P_2} = \sqrt{\frac{\pi_1(1-\pi_1)}{n_1} + \frac{\pi_2(1-\pi_2)}{n_2}}.$$

The sampling distribution of p1- p2 is approximately normal as long as the proportions are not too close to 1 or 0 and the sample sizes are not too small.

As a rule of thumb, if n1 and n_2 are both at least 10 and neither is within 0.10 of 0 or 1 then the approximation is satisfactory for most purposes. An alternative rule of thumb is that the approximation is good if both $N\pi$ and $N(1 - \pi)$ are greater than 10 for both π_1 and π_2.

The application of this sampling distribution, assume that 0.8 of high school graduates but only 0.4 of high school drop outs are able to pass a basic literacy test.

If 20 students are sampled from the population of high school graduates and 25 students are sampled from the population of high school drop outs, what is the probability that the proportion of drop outs that pass will be as high as the proportion of graduates?

For this example, the mean of the sampling distribution of p1 - p2 is:

$$\mu = \pi_1 - \pi_2 = 0.8 - 0.4 = 0.4.$$

The standard error is:

$$\sigma_{P_1-P_2} = \sqrt{\frac{\pi_1(1-\pi_1)}{n_1} + \frac{\pi_2(1-\pi_2)}{n_2}} = \sqrt{\frac{0.8(1-0.8)}{20} + \frac{0.4(1-0.4)}{25}}$$

$= 0.133.$

The solution to the problem is the probability that p1 - p2 is less than or equal to 0. The number of standard deviations above the mean associated with a difference in proportions of 0 is:

$$z = \frac{p_1 - p_2 - \mu_{p_1-p_2}}{\sigma_{p_1-p_2}} = (0-0.4)/0.133 = -3.01$$

From z table it can be determined that only 0.0013 of the time would p1 - p2 be 3.01 or more standard deviations below the mean.

ESTIMATION THEORY

Estimation theory is a branch of statistics and signal processing that deals with estimating the values of parametres based on measured/empirical data.

The parametres describe an underlying physical setting in such a way that the value of the parametres affects the distribution of the measured data. An estimator attempts to approximate the unknown parametres using the measurements.

For example, it is desired to estimate the proportion of a population of voters who will vote for a particular candidate. That proportion is the unobservable parametre; the estimate is based on a small random sample of voters.

Or, for example, in radar the goal is to estimate the location of objects (airplanes, boats, etc.) by analyzing the received echo and a possible question to be posed is"where are the airplanes?" To answer where the airplanes are, it is necessary to estimate the distance the airplanes are at from the radar station, which can provide an absolute location if the absolute location of the radar station is known.

In estimation theory, it is assumed that the desired information is embedded in a noisy signal. Noise adds uncertainty, without which the problem would be deterministic and estimation would not be needed.

ESTIMATION PROCESS

The entire purpose of estimation theory is to arrive at an estimator, and preferably an implementable one that could actually be used. The estimator takes the measured data as input and produces an estimate of the parametres.

It is also preferable to derive an estimator that optimality. Estimator optimality usually refers to achieving minimum average error over some class of estimators, for example, a minimum variance unbiased estimator. In this case, the class is the set of unbiased estimators, and the average error measure is variance (average squared error between the value of the estimate and the parametre). However, optimal estimators do not always exist.

These are the general steps to arrive at an estimator:

- In order to arrive at a desired estimator for estimating a single or multiple parametres, it is first necessary to determine a model for the system. This model should incorporate the process being modeled as well as points of uncertainty and noise. The model describes the physical scenario in which the parametres apply.
- After deciding upon a model, it is helpful to find the limitations placed upon an estimator. This limitation, for example, can be found through the Cramér-Rao bound.
- Next, an estimator needs to be developed or applied if an already known estimator is valid for the model. The estimator needs to be tested against the limitations to determine if it is an optimal estimator (if so, then no other estimator will perform better).
- Finally, experiments or simulations can be run using the estimator to test its performance.

After arriving at an estimator, real data might show that the model used to derive the estimator is incorrect, which may require repeating these steps to find a new estimator. A non-implementable or infeasible estimator may need to be scrapped and the process started anew.

In summary, the estimator estimates the parametres of a physical model based on measured data.

BASICS

To build a model, several statistical"ingredients" need to be known. These are needed to ensure the estimator has some mathematical tractability instead of being based on"good feel".

The first is a set of statistical samples taken from a random vector (RV) of size N. Put into a vector,

$$x = \begin{bmatrix} x[0] \\ x[1] \\ \vdots \\ x[N-1] \end{bmatrix}$$

Secondly, we have the corresponding M parametres,

$$G = \begin{bmatrix} \theta_1 \\ \theta_2 \\ \vdots \\ \theta_M \end{bmatrix}$$

which need to be established with their probability density function or probability mass function,

$$P(x|\theta)$$

It is also possible for the parametres themselves to have a probability distribution (e.g., Bayesian statistics). It is then necessary to define the Bayesian probability,

$$\pi(\theta).$$

After the model is formed, the goal is to estimate the parametres, commonly denoted $\hat{\theta}$, where the"hat" indicates the estimate.

One common estimator is the minimum mean squared error (MMSE) estimator, which utilizes the error between the estimated parametres and the actual value of the parametres,

$$e = \hat{\theta} - \theta$$

as the basis for optimality. This error term is then squared and minimized for the MMSE estimator.

POINT ESTIMATION

When a parametre is being estimated, the estimate can be either a single number or it can be a range of scores. When the estimate is a single number, the estimate is called a"point estimate"; when the estimate is a range of scores, the estimate is called an interval estimate. Confidence intervals are used for interval estimates.

As an example of a point estimate, assume you wanted to estimate the mean time it takes 12-year-olds to run 100 yards. The mean running time of a random sample of 12-year-olds would be an estimate of the mean running time for all 12-year-olds. Thus, the sample mean, M, would be a point estimate of the population mean, μ.

Often point estimates are used as parts of other statistical calculations. For example, a point estimate of the standard deviation is used in the calculation of a confidence interval for μ. Point estimates of parametres are often used in the formulas for significance testing.

Point estimates are not usually as informative as confidence intervals. Their importance lies in the fact that many statistical formulas are based on them.

CHARACTERISTICS OF ESTIMATORS

Statistics are used to estimate parametres. Three important attributes of statistics as estimators are covered in this text: unbiasedness, consistency, and relative efficiency.

For example, the sample mean, M, is an unbiased estimate of the population mean, μ. All statistics covered will be consistent estimators. It is hard to imagine a reasonably-chosen statistic that is not consistent.

When more than one statistic can be used to estimate a parametre, one will naturally be more efficient than the other(s). In general the relative efficiency of two statistics differs depending on the shape of the distribution of the numbers in the population. Statistics that minimize the sum of squared deviations such as the mean are generally the most efficient estimators for normal distributions but may not be for highly skewed distributions.

INTERVAL ESTIMATION

In statistics, interval estimation is the use of sample data to calculate an interval of possible (or probable) values of an unknown population parametre, in contrast to point estimation, which is a single number. Neyman (1937) identified interval estimation ("estimation by interval") as distinct from point estimation ("estimation by unique estimate"). In doing so, he recognised that then-recent work quoting results in the form of an estimate plus-or-minus a standard deviation indicated that interval estimation was actually the problem statisticians really had in mind.

The most prevalent forms of interval estimation are:

- Confidence intervals (a frequentist method); and
- Credible intervals (a Bayesian method).

Other common approaches to interval estimation, which are encompassed by statistical theory, are:

- Tolerance intervals
- Prediction intervals - used mainly in Regression Analysis

There is a third approach to statistical inference, namely fiducial inference, that also considers interval estimation. Non-statistical methods that can lead to interval estimates include fuzzy logic. An interval estimate is one type of outcome of a statistical analysis. Some other types of outcome are point estimates and decisions.

DISCUSSION

The scientific problems associated with interval estimation may be summarised as follows:

- When interval estimates are reported, they should have a commonly-held interpretation in the scientific community and more widely. In this regard, credible intervals are held to be most readily understood by the general public. Interval estimates derived from fuzzy logic have much more application-specific meanings.
- For commonly occurring situations there should be sets of standard procedures that can be used, subject to the checking and validity of any required assumptions. This applies for both confidence intervals and credible intervals.
- For more novel situations there should be guidance on how interval estimates can be formulated. In this regard confidence intervals and credible intervals have a similar standing but there are differences:
- Credible intervals can readily deal with prior information, while confidence intervals cannot.
- Confidence intervals are more flexible and can be used practically in more situations than credible intervals: one area where credible intervals suffer in comparison is in dealing with non-parametric models.

- There should be ways of testing the performance of interval estimation procedures. This arises because many such procedures involve approximations of various kinds and there is a need to check that the actual performance of a procedure is close to what is claimed. The use of stochastic simulations makes this is straightforward in the case of confidence intervals, but it is somewhat more problematic for credible intervals where prior information needs to taken properly into account. Checking of credible intervals can be done for situations representing no-prior-information but the check involves checking the long-run frequency properties of the procedures.

Severini (1991) discusses conditions under which credible intervals and confidence intervals will produce similar results, and also discusses both the coverage probabilities of credible intervals and the posterior probabilities associated with confidence intervals.

HYPOTHESIS TESTING

Whenever we have a decision to make about a population characteristic, we make a hypothesis.

Some examples are:

$$m > 3$$

or

$$m \neq 5.$$

Suppose that we want to test the hypothesis that $\mu \neq 5$. Then we can think of our opponent suggesting that = 5. We call the opponent's hypothesis the null hypothesis and write:

$$H_0: m = 5$$

and our hypothesis the alternative hypothesis and write,

$$H_1: m \neq 5$$

For the null hypothesis we always use equality, since we are comparing with a previously determined mean. For the alternative hypothesis, we have the choices: <, >, or ≠

PROCEDURES IN HYPOTHESIS TESTING

When we test a hypothesis we proceed as follows:

Formulate the null and alternative hypothesis:

- Choose a level of significance.
- Determine the sample size. (Same as confidence intervals)
- Collect data.
- Calculate z (or t) score.
- Utilize the table to determine if the z score falls within the acceptance region.
- Decide to
  - Reject the null hypothesis and therefore accept the alternative hypothesis or
  - Fail to reject the null hypothesis and therefore state that there is not enough evidence to suggest the truth of the alternative hypothesis.

ERRORS IN HYPOTHESIS TESTS

We define a type I error as the event of rejecting the null hypothesis when the null hypothesis was true. The probability of a type I error (a) is called the significance level.

We define a type II error (with probability b) as the event of failing to reject the null hypothesis when the null hypothesis was false.

Example: Suppose that you are a lawyer that is trying to establish that a company has been unfair to minorities with regard to salary increases. Suppose the mean salary increase per year is 8%.

You set the null hypothesis to be,

H0: m =.08

H1: m <.08

Q. What is a type I error?

A. We put sanctions on the company, when they were not being discriminatory.

Q. What is a type II error?

A. We allow the company to go about its discriminatory ways.

6

Economy Growth

CHARACTERISTICS

Since the early 1990s, the world has witnessed the spectacular growth of the economies of China and India (averaging 10.2 and 6.2 per cent annually from 1992 to 2005, respectively). Associated with this growth has been the dramatic development of the service sectors in the two Asian giants. However, a comparison of China's and India's economic structures demonstrates that the role of the service sector (or, the tertiary sector as it is known in China) is very different. In India, the service sector has become the dominant contributor to the Indian economy, accounting for 54.2 per cent of GDP in 2004.

The success in this sector is regarded as"India's services revolution". In China, however, the service sector has lagged well behind the manufacturing sector (or the secondary sector, according to Chinese terminology), though its role in the economy improved slightly in the last 15 years.

From 1990 to 2004, the service sector as a proportion of China's GDP increased modestly from 34.3 per cent in 1990 to 40.7 per cent in 2004. Why have the two countries taken very different trajectories in developing their service economies? What are the implications for future development in the two Asian giants? Which factors affect demand for services in China and India? These are some of the questions which are investigated in this chapter.

To date there have hardly been any comparative studies of services in China and India. However, several studies have focused on the service sector in the individual countries. For example, Gupta and Mohan both examine productivity in the Indian service sector in comparison with other Asian economies. Chanda discusses service trade in the world, particularly in India, and its implications for the World Trade Organisation (WTO) negotiations in services, while Gordon and Gupta present a detailed study explaining India's service growth in the past decade. Examples of studies on China include Li and Hou who produced an edited volume on China's WTO entry and the implications for the service sector, Jiang who edited a book focusing on growth and structural changes in services with some marginal coverage of international comparison, and Li who conducted a comprehensive investigation of China's service sector. Thus, it is the goal of this study to extend the existing literature by comparing the growth in and demand for services in China and India.

First there is a brief review of developments in the service sector in China and India. This is followed by a discussion of the determinants of the demand for services. Three empirical models are employed to examine the factors affecting demand for services internationally, as well as in China and India. Subsequently, the chapter discusses the growth outlook for services in the two countries. The final section summarises the findings.

DEVELOPMENT

Following the conventional classification, an economy is divided into three sectors, that is, agricultural (or primary), manufacturing (or secondary) and service (or tertiary). The agricultural sector consists of farming, forestry, animal husbandry and fisheries. The manufacturing sector is composed of mining, construction and manufacturing. All other economic activities which are not covered by the agricultural or manufacturing sectors are broadly defined as services and hence belong to the service sector.

They include services provided for the agricultural sector, activities associated with the supply of water, electricity and gas, transport and communications, wholesale and retail trade, finance and insurance, business and personal services, and community and social services. Services can be broadly distinguished between two types, that is, old and new. The old or traditional services include petty trading, domestic services, catering and hotel services. The new services are generally associated with communications, business and legal practices, culture, research and education. The latter are tradable internationally and hence are also called tradable services.

SERVICE SECTOR GROWTH

From 1978 to 2004, the service sectors of China and India achieved high annual growth averaging 10.8 per cent in China and 7.0 per cent in India. The difference in these growth rates matches the difference in the rates of GDP growth in the two countries. During the same period, an average annual growth rate of 9.7 per cent was recorded for the Chinese economy compared to 5.4 per cent for India. At the disaggregate level, the structures of the service sector in China and India are very similar.

In both countries, the service sector is still dominated by the traditional or old services, followed by business services (finance, insurance and real estate) and transport and communications. However, the new service sectors are catching up rapidly. From 1999 to 2003, communication services showed the strongest growth in both countries (with an average annual rate of 16.8 per cent in China and 23.9 per cent in India), followed by education (8.6 per cent) and research (11.9 per cent) in India, and real estate (8.8 per cent) and research (8.9 per cent) in China. Though growing rapidly in recent years, research-related services are still relatively small in both countries.

OUTPUT AND EMPLOYMENT SHARES

The development of the service sectors can also be examined by analysing the sectoral shares of output and employment over the national totals in the two countries.

Output shares of the service sector in China and India have grown steadily since 1978. The role of services in the Indian economy is clearly more significant than that in the Chinese economy. Accordingly, since 1990, India's service sector has grown faster (annual growth rate of 7.5 per cent) than the manufacturing sector (6.0 per cent) while the opposite is true in China where the manufacturing sector grew at 12.1 per cent and the service sector, at only 8.4 per cent.

In contrast, the employment share of India's service sector has grown very slowly according to available statistics covering the period from 1983 to 1999, though the absolute figure was much higher than that in China in the 1980s. The employment share of China's service sector has, however, been growing steadily since 1978, in particular since the early 1990s. As a result, China's service sector has absorbed relatively much more labour than India's since the mid-1990s.

In terms of creating new jobs, services have played a vital role in both economies. For example, during the decade of 1995 to 2004, more than 62 per cent of the new jobs created in China were in the service sector while India's services provided about 54 per cent of new jobs in the country from 1994 to 1999. As both countries must relocate a large pool of rural surplus labour, further employment growth in the service sector will be indispensable.

SERVICE CLASSIFICATION

REGIONAL VARIATIONS

In terms of the level of development of the service sector, there is considerable variation across the regions of both countries. "Regions" here are defined according to the official administrative units, that is, they mean China's provinces, autonomous districts and municipalities and India's union states and territories. Several observations about service development among the regions can be made. First, regional development in services follows the national trend. In general, the service sectors at the regional level in India are on an average more developed than those in China. For example,

Beijing has China's most advanced service sector, accounting for 61 per cent of GDP in 2001, while Delhi has India's most advanced service sector with a share of 77 per cent in the same year.

Second, the service sector tends to be more important among relatively more developed regions in both countries. There is clearly a positive relationship between the level of service sector development and GDP per capita in both Chinese and Indian regions. Finally, the service sector plays a more important role in regions where the level of urbanisation is high, such as in Delhi and Chandigarh in India, and Beijing and Shanghai in China.

GROWTH IN INTERNATIONAL PERSPECTIVE

The experience of economic development shows that when a country expands its manufacturing capacity, its agricultural sector declines, and that over time the service sector grows and eventually overtakes the manufacturing sector to become dominant. For example, the share of services in total GDP in 2003 amounted to 71 per cent in Australia, 72 per cent in the United Kingdom and 75 per cent in the United States. This process of change follows the stages of economic development.

However, in comparison with countries at a similar stage of development, China's service sector development is slightly below the average while India's is slightly above it. These are clearly demonstrated which illustrates service sector GDP shares against per capita income among nations with per capita GDP less than USD2000 in 2003. Information at the disaggregate level also implies that services in China and India are still at the early stage of development which is represented by the dominance of the old services, as well as weakness in real estate, education and research services.

The relative backwardness of services in China is also reflected in the role of service trade. In 2003, China was the world's fourth largest merchandise exporter but its service exports ranked only tenth. The share of service exports over total exports in China (9.6 per cent in 2003) was much smaller than the 28.4 per cent in the United States, 32.4 per cent in the United Kingdom, 22.4 per cent in Australia, 13.9 per cent in Japan and 18.6 per cent in the world as a whole during the same period. Not surprisingly, China is currently a net importer of services.

In contrast, India's service exports accounted for 30.9 per cent of total exports in 2003 making the country a net exporter of services according to the World Bank. This is largely due to India's success in information technology (IT) service exports and India's relatively small value of merchandise exports (equivalent to only 12 per cent of China's in 2003). In 2003, India's total merchandise exports ranked 31st in the world but its service exports ranked 20th.

In addition, among the service exports in 2003, the share of finance and insurance amounted to only 1.0 and 1.5 per cent in China and India, respectively, while this share was 7.8 per cent in the United States and 22.6 per cent in the United Kingdom in the same year according to the World Bank.

To sum up, the service sector as part of the national economies of China and India achieved substantial growth in recent decades. However, from an international perspective, China and India are outliers, i.e., India's service sector over-performed while China's under-performed. In addition, new services are still weak in both China and India with the exception of India's IT export sector. At the regional level in each country, the pattern of service development is similar to the national trend. In particular, service development seems to be closely related to the level of urbanisation. These features are further explored in the following sections.

It is argued that service sector growth is determined by several factors such as production specialisation, income level and urbanisation.

These factors are interrelated. As an economy grows, productive activities become more specialised and urbanisation accelerates due to the rising level of income. In the meantime, as a result of the increasing specialisation of production, firms tend to outsource many service activities such as legal, accounting and security services. Some authors call this process "specialisation splintering". The main source of demand for services comes from producers.

At the household level, Engel's law shows that, as income rises, consumption of food and durable goods becomes saturated over time and demand for services such as healthcare, travel, communications and finance increases. Growth in income also boosts demand for away-from-home consumption of food and services. Furthermore, urbanisation contributes to the growth of the service sector in two ways. Unlike farmers who, to some extent, are self-sufficient in service provision, urban consumers rely on commercial suppliers to provide most services. They are also more likely to enter the urban informal sector for employment if there are no job opportunities in the formal sector.

Services account for the lion's share of the informal sector. Thus, urban residents are both consumers and suppliers of services. It is expected that the service sector grows as the level of urbanisation increases. This is confirmed by preliminary analysis of both China's and India's regional data in the preceding section. In addition, the participation of women in the workforce has an impact on service demand as well. More women in the workforce can lead to an increase in demand for services, ranging from babysitting and catering to tuition and beauty treatments. Furthermore, regulatory policies also affect the development of the service sector. A good example is the rapid growth of telecommunications services after deregulation in many countries.

This phenomenon can also occur in other areas such as insurance, banking, health care, etc. Regulatory environments can also affect international trade and foreign investment in services. Finally, apart from the factors discussed above, service sector growth in China and India may also be affected by development strategy and historical experience. Divergence in the latter may contribute to the different role of services in the two countries. For example, India's service industry may have benefitted from the country's early and prolonged linkage with the West, in particular the UK. In the meantime, when the communist government took office in China in 1949, it adopted a biased policy towards the development of heavy industry initially, and labour-intensive manufacturing more recently, while the movement of rural-urban labour was restricted for some four decades and hence entry into the service sector (mainly old services) was very limited. Thus, China's historical development policy was anti-services and this partly explains the underdevelopment of the country's service industry. In the empirical analysis presented in the following section, the factors just discussed are taken into consideration to the extent that quantification is possible.

THE EMPIRICS

To examine the determinants of growth in services in China and India, three models are applied to Chinese provincial, Indian state and cross country databases, respectively. For panel data models, several optional models are considered, e.g., without group dummy (the same intercept for all groups), fixed effect (different intercept for each group) and random effect models (intercepts vary by a random error).

SER in equation (1) is measured as the GDP share of services in a country or region. The choice of the variables (X) in the models is dictated by the availability of data. In the three optional models considered here, X includes per capita income (I), urbanisation (U), service export shares in total exports (EX) and the proportion of women in the total non-farming workforce (W). EX and W are not available for the regions of China and India, and are therefore incorporated in the cross-country model only. The Chinese data cover 31 regions and the period 1993 to 2003. The Indian statistics are available for 31 union states and territories out of 35 for nine years. The cross-country database has one year (2003) data for 93 countries.

The estimated coefficients of both per capita income and urbanisation are positive and statistically significant, implying that both variables are important factors affecting the development of the service sector, especially in China and India. In addition, it is found that external demand also has a positive impact on service development among the countries.

This partly explains the phenomenal growth of IT service exports in India. The cost of services is shown to have a negative effect but the estimated coefficient is statistically insignificant. Thus, further investigation is needed. The findings here are of course subject to qualifications due to data limitations.

SERVICE MARKETING

GROWTH OUTLOOK

Both the Chinese and Indian economies have entered a phase of rapid growth. It is commonly agreed that this growth will last for decades. It is also anticipated that per capita income in both countries will rise over time and that urbanisation will subsequently accelerate. These factors imply that the Chinese and Indian service sectors will expand further. This growth will also be boosted by globalisation. From 2006, China's WTO commitments will allow greater foreign participation in the service sector including telecommunications, banking, insurance, etc.

Economic openness will also create more jobs for accountants, lawyers and other financial specialists. Further economic liberalisation and deregulation in India will ensure sustained growth, including growth in services. Both China and India will also benefit from the WTO's General Agreement on Trade in Services (GATS). Therefore, the growth prospect for services in both countries is bright and will be associated with the following features.

Service sector growth in China will continue while the manufacturing sector remains dominant. Both sectors will show some gain in terms of both GDP and employment shares at the expense of agriculture. According to one set of forecasts, the GDP shares of the agricultural, manufacturing and service sectors in 2020 will be 5.0, 45.8 and 49.2 per cent, respectively, while the employment shares will be 34.2, 22.4 and 43.4 per cent, respectively. By then, China's service sector will be larger than the manufacturing sector. However, China will still have a long way to catch up with the advanced economies in the world. For instance, the service sector in 1997 accounted for 58 per cent of total GDP in Japan, 64 per cent in the US, 53 per cent in Germany and 60 per cent in the UK, while the employment shares were, respectively, 62 per cent in Japan, 74 per cent in the US, 62 per cent in Germany and 72 per cent in the UK. In China, the service sector contributes to national economic growth largely through the shift of labour from agriculture into services. This shift of labour will eventually be replaced by the shift from manufacturing to services.

The current growth momentum in India's service sector will continue in the near future. However, growth in the manufacturing sector will be accelerated. This trend is already clearly demonstrated in the latest statistics. From 1994 to 2003, the manufacturing sector grew at an average annual rate of 6.6 per cent, though the growth rate increased to 8.1 and 9.0 per cent during the financial years of 2004-5 and 2005-6, respectively.

The consequence of such growth is that the GDP shares of both the manufacturing and service sectors will expand at the cost of agriculture. For example, the GDP share of services was estimated to be 57.6 per cent by the end of the 2004-5 financial year.

The dilemma for India at present is that services account for more than a half of GDP but employ less than 30 per cent of the total work force. According to the experience of developed economies as cited in the preceding section, these shares should match. A study of the Pacific Basin countries also shows that without concomitant progress in industrialisation, growth in services in terms of both output and employment is not sustainable. The same argument can be applied to India as well.

In addition, recent expansion of India's service sector, to some extent, is boosted by the IT service exports but, the IT sector employs mainly educated, urban youth, potentially leaving most of India's population further behind. Furthermore, the numbers employed in India's IT sector are relatively small (under one million) and Konana et al. argue that its rapid growth has not generated sufficient linkages with the rest of the economy. Thus, India will need growth in both services and manufacturing.

It is found that the role of services in both China and India has been rising, with China starting from a lower base. Growth has been driven in both societies mainly by increasing specialisation of production, rising living standards and accelerated urbanisation. There are also some non-economic factors which are difficult to quantify in empirical analysis but have played important roles in service development, including biased development strategies in China, India's early linkage with the West and the recent boom in Indian IT exports.

India's service sector is now the dominant contributor to GDP growth but employment absorption is not very high. India's service sector will continue to grow, but the country needs industrialisation even more as millions of rural workers have to be employed. The IT sector is not large enough to significantly affect the growth of the national economy. In comparison with India, China's service sector is lagging behind. Even by international standards, China's service sector is below average and it can be anticipated that services will expand further as Chinese companies outsource their communications, legal and accounting services and as urbanisation accelerates in the coming decades. In addition, the service sector has been the main provider of new jobs in China where there still exists a sizable pool of surplus rural labour.

The first draft was completed while the author was visiting the East Asian Institute, National University of Singapore. The author is grateful to its Director Professor Wang Gungwu, and Research Director Professor John Wong for the generous support provided. He also thanks three anonymous referees, as well as Elspeth Thomson and participants at an EAI seminar for helpful comments and suggestions.

QUALITY GAPS MODEL

Since the end of World War II, jobs in the service sector have steadily increased from a little less than half of all jobs in the U.S. economy to nearly 80%. One of the most important industries found within the service sector is transportation, which in recent years has accounted for over 20% of the nation's gross national product (GNP).

The importance of service quality in any service industry cannot be disputed. Recent political, economic, and technological changes affecting the transportation industry have made service quality a major concern for carriers and shippers alike. Shippers have increased expectations concerning the quality of service they receive and carriers are struggling to meet these expectations. This struggle between shippers and carriers would suggest that there is room for improvement in carrier managements' understanding of how shippers define service quality. This study is an attempt to analyse service quality within the industry and to determine potential areas of improvement within the carrier/shipper relationship.

DEFINING SERVICE QUALITY

A recent Gallup survey indicated that executives ranked the improvement of service and tangible product quality as the single most critical issue facing U.S. managers. But how do we define service quality Zeithaml, Parasuraman, and Berry tried to answer this question as early as 1983. Their research showed that service quality can be defined as the extent of discrepancy between customers' expectations or desires of service and their perceptions of the service they actually receive. These same researchers have developed a model of service quality that focuses on possible communication and control gaps within the "consumer-marketer" system, which can lead to a breakdown in service quality. These gaps have been tested and supported in earlier research using several types of service organizations. Since transportation is such a vital segment of the U.S. service economy, it would seem fruitful to examine the industry to discover whether it suffers from a breakdown in service quality. This study will assess these service gaps and offer solutions to any service quality problems.

SERVICE QUALITY MODEL

The service quality model, as adapted to the transportation industry. The first gap represents differences in expectations between shippers and carriers. This gap indicates the amount of discrepancy between what service shippers actually expect and the service carriers think shippers expect.

Shippers may be saying to carriers, "We want on-time delivery" and carriers may be hearing "We don't want any late deliveries." In this case, there could be a significant difference between what the shippers are saying and what the carriers are saying. If carriers really understand and communicate well with their customers, misunderstandings should seldom occur, and gap one should be quite small.

The second gap illustrated in the model is found only on the carrier side. This gap occurs when management incorrectly translates shipper expectations into service quality specifications. For instance, management may define on-time delivery as any shipment arriving within four hours of the scheduled arrival, while the shipper may expect delivery as scheduled plus or minus two hours.

The third gap is caused when service delivery does not meet service quality specifications. This gap is also found only on the carrier side of the model and represents a breakdown in customer service implementation within the organization. According to Zeithaml, Parasuraman, and Berry, this can be caused by a lack of teamwork (employees and management not working towards a common goal), or confusion surrounding control (who is accountable for quality control). Managers must communicate standards and quality specifications clearly to employees and reward those who implement them.

For ease of analysis, gaps two and three have been condensed into one gap, which will measure the difference between what the carrier believes the shipper wants and what the carrier actually delivers. Should this study reveal a significant gap 2/3, it will be necessary to examine both gaps separately in future studies.

Gap four of the service model is the difference between the service delivered and the external communications from the carrier to the shipper.

For example, if the carrier sales representative promises 90% on-time delivery (external communication), and the carrier delivers only 80% on-time, this difference will appear in the fourth gap.

The two main causes of this gap, according to the model's authors, are lack of horizontal communication (i.e., a breakdown in communication between sales and operations personnel), and the propensity to overpromise (i.e., the sales force making promises to shippers that they know cannot be met). Sometimes salespeople misunderstand the fact that unless operations can deliver on their promises, the entire organization suffers.

Gap five says there is a significant difference between the customers' expectations of the service quality and the quality of service the customer receives. This gap is the direct and indirect result of adverse effects in the four prior gaps in the model. It measures the cumulative impact of the differences in any or all of the first four gaps.

Based on this service quality model, the following hypotheses will be tested in this study:

- *H sub* 1: The service that shippers expect is significantly different from the service carriers believe shippers expect. (Gap 1)
- *H sub* 2: The service that carriers actually deliver is significantly different from the service carriers believe shippers expect. (Gaps 2/3)
- *H sub* 3: The service that carriers actually deliver is significantly different from the service promised to shippers. (Gap 4)
- *H sub* 4: The service perceived by the shippers is significantly different from the service shippers expect from carriers. (Gap 5)

According to the service quality model, if hypothesis one, two, or three is supported by the study, indicating gaps in the shipper/carrier system, hypothesis four will also be supported.

If there are no gaps in the system, then shippers are receiving the service quality that they expect.

Servqual, a multiple-item scale for measuring perceptions of service quality, was the survey instrument used in this study. This scale, which was developed and tested by Parasuraman, Xeithaml, and Berry, was found to have Total-Scale Reliability coefficients ranging between.87 and.90 for the various dimensions of service quality. Details about how the survey instrument was developed and tested can be found in the article cited in the reference section of this chapter. Questions from the survey were used to test the four hypotheses formulated for this study. Each question uses a seven-point, Likert-type scale, ranging from strongly agrees to strongly disagree. The survey was mailed to attendees of a national transportation educational seminar, and members of a national transportation professional organization.

The sample could be considered a "quasi" random sample. It represents a "cluster sample" or cross-section of the industry, and includes representatives of most modal classes, as well as shippers who use various modes within the industry. This method of sampling was used due to the prohibitive cost of obtaining a listing of all carriers in the industry and a comparable listing of all the shippers who utilize the services of these carriers. Of the 950 surveys mailed, 148 usable responses were returned. The 66 respondents employed by carriers and 82 respondents employed by shippers represent a response rate of nearly 16%.

The respondents are representative and knowledgeable about the industry. The carrier respondents have been involved in transportation for an average of 17 years (modal response 10 years) and the shippers have been involved an average of 15 years (modal response 20 years). Job titles reflected in this sample include president/vice president, director, manager, supervisor, and a general category called other.

Transportation modes represented by both carrier and shipper include air cargo, motor carrier (general commodities and household), rail, and sea cargo. The questionnaire was based on the reactions of shippers and carriers to various statements reflecting features of service quality. These statements are based on a survey developed by Zeithaml, Parasuraman, and Berry. The authors describe twenty-two statements that were designed to measure service-provider gaps and their causes. Statements applicable to transportation were selected and adapted to specifically analyse possible gaps in the transportation industry. Data from the questionnaires provided information on what the service shippers expect, carriers perceptions of their shippers' service expectations, the actual service delivered, external communications to shippers, and the service shippers actually perceived that they received.

Dependent and independent t-tests were performed to test the hypothesized differences and to examine the gaps as explained in the service quality model. The results of these tests and their conclusions will be discussed in the remaining portions of this chapter.

To measure the existence of the first gap of the model, shippers were asked to rate what they believe to be important features of an excellent carrier (expected service quality) and carriers were asked to rate the features the carriers were actually providing the shippers (perceptions of expected service). Respondents were asked to assign a number from one (strongly disagree) to seven (strongly agree) for each statement, representing the degree of importance they attach to each feature of service quality.

It might seem logical that, initially, one would expect differences to exist between what the shipper expects and how the carrier understands these expectations. However, the service model dictates that if the carrier is to provide excellent service, it must clearly understand the customer's (shipper's) expectations and then provide service consistent with these expectations. Therefore, we would not expect significant differences between shipper expectations and how the carrier understands these expectations when excellent service is being provided.

It is interesting to note that only six statements showed the expected and perceived results to be significantly different (alpha <.05). Modern equipment, personal and individual attention given to shippers, error-free records, and convenient operating hours were all judged to be more important by carriers than by the shippers. Carriers' delivery on promises was the only statement judged significantly higher by shippers than carriers. This finding emphasizes how important promises are to shippers.

These results may indicate that the first gap in the model is a function of carriers no understanding the importance shippers place on promises. If true, this should be reflected in the remaining four gaps in the model.

The data supports the first hypothesis—that there is a significant difference in expected and perceived service quality. However, this difference is only apparent in six of the nineteen statements, and in only one is the carrier underestimating the importance of a feature of service quality.

GAP TWO/THREE

Gap 2/3 is measured as the difference between what the carriers believe they are actually providing (perceptions of carriers), and what the shippers believe they are actually receiving in service quality (service delivery). However, the shippers are asked not what they want, but what they are actually receiving from their primary carriers. The differences between shipper and carrier responses appear much greater in gap 2/3 than they appeared to be in gap 1.

Not only are responses to all but three of the statements significantly different, the difference favours carriers with the higher mean score in all cases, in all of the statements, the carriers overestimate the quality of service they are providing to the shippers as compared to the shippers' estimate of that same service. It is interesting to note that while the carriers were much closer to the perceptions of what the shippers expect, what the carriers provide is much further from the target. This may be explained in terms of the service quality model as the size of gap 2/3 being much larger than that of gap 1. If this is true, it would indicate that improvement should focus on translating perceptions into quality specifications and standards, then making sure these quality specifications and standards are met. These data strongly support the second hypothesis of this study: the service quality carriers actually provide is not the same as the service quality carriers believe that shippers expect from them.

GAP FOUR

Gap four of the service quality model is the difference between service delivery and external communications. One of the major components of external communications is the making of promises; carrier sales representatives me promises to their shipper customers. Therefore, this gap is measured as the difference between how both the carriers and the shippers judge how well the carriers keep those promises.

There is a significant difference in both comparisons, particularly the comparison of shippers' expectations versus their actual experiences. It would appear that the data supports the third hypothesis of this study: the service carriers actually deliver is not the same as the service promised.

GAP FIVE

Since all four hypotheses have been supported by the analysed data, it would follow that Gap 5 should also be confirmed by the data. The last gap reflects the cumulative effect of the first four gaps, and represents the difference between the service quality the shippers expect and the service quality the shippers actually receive from the carriers. The size of this gap is critical because it is this Fifth gap that forms the basis of customer dissatisfaction. It is in this fifth gap that the effects of all of the previous gaps are revealed to the shipper in the form of unfulfilled expectations.

To measure the size of Gap 5, the same statements were again given to the shippers. This time the respondents were asked to assess each statement in terms of the service they would expect from an excellent carrier and then in terms of the actual service they receive from their primary carrier.

All but one of the statements show significant differences between what the shippers want and what they believe they are receiving. Convenient operating hours is the only feature of service quality in which expectation and perception are approximately equivalent. Modern equipment shows a significant difference. However, the negative mean difference implies that the perceived service is actually superior to the expectation.

Results of the remaining seventeen of nineteen statements indicate that shippers' expectations of service quality are higher than the perceived service they are getting from their carriers. This would imply a large gap between service expectation and service receipt, and could arguably be the basis for customer service problems within the transportation industry.

The fourth and final hypothesis of this study was also strongly supported by the data; the service perceived by the shippers is not the same as the service quality expected by the shippers. As the previous three hypotheses were all supported, it follows that support for the final hypothesis would also be found.

The statement with the largest mean difference expectation and perception is "carrier delivers on promises." This may suggest that carrier sales representatives who make promises to shippers that cannot be kept are the major source of shipper dissatisfaction.

Based on the results of the analyses, support was found for all of the four hypotheses proposed in this study. In terms of the service quality model, the data also confirmed the existence of all five theorized gaps in the carrier/shipper system within the transportation industry.

Identifying and understanding the gaps in the carrier-shipper relationship will be a firs step in allowing individual carriers to focus their attention on areas specifically needing attention. Results of this study suggest that carriers understand what shippers expect from a carrier. This implies good communication between the shippers and carriers, a definite strength for the industry.

However, there seems to be a breakdown somewhere on the carrier side. The carriers know what service quality the shippers expect, and yet it would seem that the carriers are not providing that quality of service to the shippers.

Why is his happening? Do the carriers feel it is too costly, or are they failing to translate what they know of shippers' expectations into measurable quality standards and specifications that their employees can follow? Or are carriers no communicating standards and specifications clearly to their employees?

These questions remain to be answered by the individual carrier. The next step in this stream of research is to examine the three gaps to determine more precisely where and why the breakdowns ate occurring.

The size of Gap 5 is consistent with the literature that states that shipper expectations of service quality are not being met. This research suggests that the gap includes many features of service quality that appear not to meet expectations and points to many areas in need of improvement. The most conspicuous offender is in the area of promises and lack of delivery on those promises. By concentrating on keeping their promises, carriers may be able to improve the service quality they offer their shippers.

Because the keeping of promises is important in shaping shipper expectations, a small improvement in this area could make a significant difference in the shippers' perception of service quality. Focusing on this one area may afford carriers dramatic results in their effort to improve service quality. This in turn may translate into a significant positive impact on profits.

Several other features of service quality bear investigation. The data indicate that modern equipment is overrated by the carriers, and is of less interest to the shipper. To better meet shipper expectations, carriers should emphasize other service features in lieu of modern equipment.

Other areas in need of improvement appear to be performing service on time, being more interested in shipper's problems, and error reduction in records. Again, future studies should focus on each of these areas to discover the fastest and most efficient methods for improvement.

SERVICE MARKETING MIX

It is important to note that this study assumes that the respondents are characteristic of the industry as whole, individual shippers and carriers may vary in how well they conform to these results. Nonetheless, it is useful for all carriers to address these issues within their organizations to ensure that the service quality they provide is the best that they can possibly provide, given their individual constraints.

The results are useful to individual carriers in several ways. First, it is hoped that the results will prompt individual carriers to constantly monitor their customers' (shippers') expectations and make sure they know how well they are being met. One way to do this would be to administer the Servqual instrument regularly or develop a questionnaire. The results from the questionnaire could be used in conjunction with customer comments and other sources of information to tell the carrier how well it is doing.

Second, the results of this study should motivate the individual carrier to determine how well its representatives keep the promises they make to shippers. This may entail developing new performance measures to evaluate carrier representative effectiveness. The literature emphasizes how critical keeping promises are to customer satisfaction.

Lastly, the results of this study indicate that from an industry vantage point, carriers can always try to improve the service they provide. As we currently find ourselves in the midst of a significant downturn, excellent service can make the difference in a carrier's survival. Whether the carrier is a trucking company or air freight company or any model, the principles embodied in the service quality model apply to each and every mode. Although the results are combined to reflect the state of service quality in the industry, it must be remembered that individual carriers and modes make up the industry.

By concentrating on the problems disclosed in this study, and evaluating their performance relative to the findings, individual carriers should be able to discover ways to improve their service quality. Findings of future research should help to pinpoint problem areas better and to offer definitive solutions to the improvement of service quality in the transportation industry. This in turn will enhance the profitability of the industry as a whole.

Quality may be the most critical component in satisfying an organization's customer. The elevation of quality as a component of customer satisfaction started with the teachings of Juran, Crosby, Deming and Feigenbaum and continued with the development of programmes such as Total Quality, Continuous Quality Improvement and Total Quality Management. Total Quality Management positioned quality as the "meeting or exceeding of a customer's expectations". Further, Garvin and Parasuraman, Zeithaml and Berry put forth the proposition that while quality was multidimensional in nature, it could be enhanced or lost unidimensionally. While this continued focus on quality contributed to a general improvement in products and services, it did not encompass all phases of a business's competitive position. Quality programmes helped define which dimensions contributed to a customer's initial perception of quality, however, they did not identify how and whether t he importance of those dimensions could vary; from the customer's initial interest in the provider's product through the product's useful life, i.e. the transaction cycle.

In order to successfully compete in the unending race for a customer's business, in an economic manner, an organization and its management must develop a competitive edge. This edge may come in the form of understanding how a company's customers value a given quality dimension (relative to the other quality dimensions) and when and if that quality dimension can increase (or decrease) in importance over time. Therefore, to clarify understanding of these quality dimensions and their interrelationships, a comprehensive model would provide operational guidelines in these competitive struggles.

In 1987, Garvin proposed a framework to describe overall product quality that consisted of eight separate dimensions. In 1988, Parasuraman, Zeithaml and Berry identified five generalizable factors (dimensions of quality) for service industries. These quality dimensions became the foundation on which much current quality dimensional research has been founded.

A review of current literature identifies a continued separation of product and service quality dimension research, an interest in service quality gaps, the relevance of these dimensions in various industries and industry-specific interpretations of these dimensions. Unfortunately, very little research has been conducted to incorporate these two sets of quality dimensions and, most importantly, no work has been done to determine the effect timing would have on the relative importance of these quality dimensions.

If timing does affect the desirability of these dimensions, then the current quality models are incomplete. This chapter is an attempt to highlight the importance of managing this comprehensive set of quality dimensions over time.

This chapter will integrate the generally accepted dimensions of quality into a general model, which can guide future research related to the interrelationships attributable to a typical transaction cycle. A general model encompassing quality dimensions and their relative importance within the transaction cycle would offer value to both academicians and practitioners. As a last step, a quantitative methodology is offered to operationalize and/or assist in the management of quality.

DIMENSIONS OF QUALITY

Quality has traditionally been viewed as: 1) conformance to requirements; 2) fitness for use; and 3) innate excellence. In 1987, Garvin suggested that product quality is not a single recognizable characteristic; rather, it is a multifaceted characteristic that appears in many forms. He observed eight dimensions of product quality: performance, features, reliability, conformance, durability, serviceability, aesthetics and perceived quality. He also maintained that different users would require different mixes (combinations of varied amounts) of the quality dimensions. In other words, quality is in the eye of the beholder.

In 1988, Parasuraman, Zeithaml and Berry identified five quality dimensions for service industries: tangibles, reliability, responsiveness, assurance, and empathy. These dimensions and the SERVQUAL instrument, from which they were derived, have received considerable support from Carman and Fick and Ritchie and some criticism from Crosby and LeMay.

Crosby and LeMay argued that price should have been added to the quality dimensions derived from the SERVQUAL instrument, thereby forcing choice and stability.

Few transactions can be identified as purely product (no service involved) or purely service (no physical product involved). Rather, most transactions provide a combination of product with accompanying service or service with some product. Goods composed of a combination of both product and service require an evaluation of both product and service quality dimensions. It, therefore, seems reasonable to merge the lists of quality dimensions from product- and service-based research.

To assist in operationalizing this dual list of quality dimensions we are suggesting a time- based approach. We, therefore, are grouping the dimensions by stages in the transaction cycle; i.e., distinguishing between those established at the time of exchange (E) and those added throughout the relationship. The transaction cycle for a product and/or service traditionally starts with the perception of demand (customer needs), moves through product/service design, manufacture and/or provision, transaction and/or sales, use of product (including support services) over its useful life and, finally, through the thought processes that may or may not lead to a repeat purchase from the same source.

The first group of quality dimensions would be established upon delivery (embedded, as it were), while the second group supports the package of delivered product/ services over the usage period in the transaction cycle (supporting). Finally, price, an estimation of the value of all attributes, will cause the customer to choose a level of each quality attribute they need or want, but not more than they want.

By including price into the model, a major problem encountered during previous uses of the SERVQUAL questionnaire would be eliminated. The instability introduced into the SERYQUAL factors, due to their desirability and non-cost, would be reduced by forcing choice. Price causes the buyer to make tradeoffs between the various quality attributes and money.

THE GENERAL MODEL

As put forward in this configuration, three components - those embedded, those supporting and their price -- contribute to customer satisfaction. The provider builds the embedded dimensions into the product before the exchange. This component is divided into two subsets of dimensions.

The first subset -- performance, features, conformance, serviceability, aesthetics and perceived quality -- can be determined at the time of exchange and cannot be augmented after the exchange. The second subset -- reliability and durability -- can only be fully established (measured) after waiting long enough to allow the purchaser to evaluate the product's performance.

The second component, supporting dimensions, is explicitly or implicitly part of the transaction.

It consists of the package of dimensions that:

- Supports the exchange process itself.
- Supports the customer's use of the goods over their life, as covered by the initial transaction.

This set of dimensions may include the appearance of any facilities where the sale and any support services take place (tangibles), willingness to help customers (responsiveness), knowledge and courtesy (assurance), and apparent degree of caring about the customers (empathy). This set of dimensions can and probably will change over the usage portion of the transaction cycle. The value of this set of dimensions can be increased, held the same, or decreased according to the seller's desires and capabilities.

Price, the third component in this model, is established at or before the transaction, representing the product/service's expected value. Whether the price fairly represents the value received by the buyer cannot be ascertained until much later in the transaction cycle, when value received is compared to value expected. The buyer cannot easily modify price after the exchange, so it is established at the time of the transaction, but value is not.

DIMENSIONS OF QUALITY

At the time of exchange information concerning features, aesthetics, perceived quality and tangibles are observable while performance, conformance and serviceability are available with additional effort on the part of the customer. However, reliability and durability are largely unknown for the specific product, but can be identified for the category of product.

Experience with the product or service is required to establish the extent of those latter dimensions. Further, responsiveness, assurance and empathy are generally available but are still under the on-going control of the seller, and therefore are subject to change over time.

The"Customer's Perception of Quality Dimensions over Time" was compiled by a focus group of several academics/ consultants to industry. The level of each dimension's importance and the information available to the customer were determined based on the focus group members' many years of experience. A series of empirical research projects are necessary to determine if these findings are generally held to be true by companies in one or more industrial SEC codes.

Further evaluations of these patterns and their availability would afford insight into the customer's level of satisfaction with its product or service purchases over the expected period of usage. A service provider may be chosen based on evident embedded dimensions, promised supporting dimensions and the price charged. However, repeat business comes from satisfaction

SATISFACTION

The desirability/importance of quality dimensions could

be depicted from the time of exchange to the end of the product/service usability. Each of the three groups of quality dimensions can be classified respectively by: dimensions embedded in the product/service, support dimensions for the product/service, and price. The correct combination and/or level of these three groups of quality dimensions would contribute vastly to any particular customer's degree of satisfaction derived from the transaction.

Certain EB1 dimensions (performance, features, conformance, serviceability and aesthetics) will be prime contributors to customer satisfaction at the time of the transaction due to the immediacy of sight, feel, sound, smell and taste of the product/service itself.

These will diminish as the product loses its feel of newness. EB2 will become more important to the customer's satisfaction as the positive or negative value of the product's reliability and durability play out. The second component (support), depicted by (SP), can increase or decrease in importance due to the continuing presence or absence of the supporting quality dimensions.

QUANTITATIVE METHODOLOGY

Traditionally, managers of the selling firm tend to view quality as inherent (embedded) in the product and/or service being sold. This leads to the mind-set that quality has been established when the product and/or service has been delivered to the buyer. Therefore, if the product did not live up to the advertised, expected levels of quality, the buyer was stuck with it and the seller's reputation "takes the hit".

This chapter takes the position that some of the dimensions of quality are still under the control of the seller and therefore can still be augmented after the point of sale or delivery.

If some of these embedded dimensions do not live up to their billing, then other dimensions can be modified or enhanced to preserve the perception of overall quality of the product and the provider. For example, if an auto manufacturer found that an unusually large percentage (i.e.,

5% versus 1%) of their transmissions developed a significant problem that required those transmissions to be replaced within 75,000 miles of use, the auto manufacturer could do one of four things:

- Do nothing;
- Test all transmissions currently in stock before installing (if possible);
- Remove all the transmissions in stock from the assembly process;
- Enhance the warranty on all transmissions.

The first option would create a frustrated, angry customer that most likely would not become a repeat customer for the auto manufacturer.

The second and third options would be very expensive, and while the cost of these actions could be passed along to the customer, it may place the auto manufacturer at a price disadvantage. The fourth option would be the best for the auto manufacturer as it would require action only on those transmissions that failed and would preserve their quality reputation while the problem was researched, resolved and incorporated into the transmission's design.

This quick, no-questions-asked service, would create a customer that is confident in the auto manufacturer's ability to handle future problems (they would be WOWED!). The enhanced warranty would represent a modification of a controllable dimension of quality that would allow the auto manufacturer to preserve the customer's perception of the overall quality of the purchased product (automobile) and the provider (auto manufacturer).

7

Perception and Expectation

Franchising, as a form of market organization, has experienced rapid growth over the last several decades. Both the franchisor and the franchisee can improve performance from a marketing perspective and a profitability context when there is an effective franchising arrangement. Market advantage can be achieved when the franchise system successfully develops an identifiable product or service which assures customers of a uniform and predictable level of product quality and service. The strength of a franchise operation is focused around the dual roles of:

- Excellent management and operations
- Outstanding quality and reliability of the product and service.

A franchisor must therefore be concerned with both the quality of its franchisees and its products.

The franchisor faces a dichotomous problem of marketing to both:

- Perspective franchisees
- End consumers or customers.

In order to grow, the franchisor must recruit additional franchisees. Effective recruitment calls for a market programme targeting perspective franchisees. In addition, the franchisor often has the responsibility of promoting the product or service to the end consumer.

The advantages inherent in franchising systems are often recognized by scholars and practitioners alike. The success that franchising has achieved during the past several decades attests to the merits of franchising. Nevertheless, there still exist within franchise relationships classic problems of control and self-interest. The conflict that arises between franchisors and franchisees may be one of the most difficult and frustrating challenges in franchising. There are often several sources of conflict within franchise channels.

Franchisees may not provide an expected level of managerial talent and energy to properly handle the business. The franchisee may be buying a vision of the American dream with unrealistic expectations of the work necessary to make the business succeed. The franchisee may not desire the hard work or long hours that are expected at the beginning of a franchising operation. The franchisee may be motivated to purchase a franchise to escape from either unemployment or poor job prospects but may not have the skills to manage his or her own business. The franchisee may not have thoroughly investigated the franchisor nor may he or she have sufficient operating capital to start up or run the business. When problems arise, the franchisee's disenchantment may seriously impair the relationship with the franchisor and cause serious harm to the business operation.

Conflict arises when one party believes that its goal attainment is being impeded by another. In a franchising relationship, the franchisor and franchisee usually have different objectives. The franchisor generally will be considering performance of the entire system while the franchisees are focusing on their individual outlets. Both franchisor and franchisee will attempt to maximize profits for themselves.

As each organization within the franchise system strives to meet its own goals, the problem of self-interest can create serious difficulties for the harmonious operation of the franchise relationship. One problem which must be addressed is the responsibility of the franchisor to properly select those franchisees whose goals complement their franchising system. Training can also be provided on an ongoing basis to ensure that the franchisees' goals remain consistent with those of the franchisor. In both cases, market research can prove to be a valuable tool to understanding franchisees' goals and aspirations.

When a suitable franchisee is enlisted, the franchisor must continue to monitor and analyse the sales and profits of each unit. Such research can provide insight into the customers' buying habits and the demand for the product or service being offered. Of course, franchisors need to adequately understand the end consumers and their desire for the franchisors' products in order to ensure their own survival.

But even more important, the franchisor can collect information which can identify and define the marketing opportunities and problems for a particular business. The ability of the franchisor to understand demand from a total systems perspective yields greater coordination of franchisor/ franchisee effort, greater capacity to take advantage of economic efficiencies in the research process, and more openness and adaptiveness to change on the part of franchisees.

The decision by franchisors to conduct marketing research is generally centreed around a recognized need for specific information and the desire to manage and maintain proper organizational functions. It is generally not thought of as a tool to enhance franchisor/franchisee relationships. It is in this area that marketing research shows great promise. To be effective, however, marketing research should address marketing issues in both target markets. First, it should help the franchisor choose franchisees whose goals will enhance franchise system performance, and second, it should help the franchisee better serve its end consumer.

THE MARKETING RESEARCH PROCESS

There are basically four steps in the marketing research process that franchisors must follow:

- Define the problem and research objectives.
- Develop a research plan for collecting information.
- Implement the research plan by collecting and analyzing the data.
- Interpret and report the findings.

When defining the problem and research objectives, it is important that the franchisors doing the research work closely with the decision makers to determine what research is needed and how it will be used.

If this step is skipped, the results generally will not address the major problem. In a franchise setting, the research people must work closely with those people in the franchise who handle recruitment of potential franchisees. They must also work with the franchisees themselves to understand the needs of the market.

Marketing research is designed to reduce uncertainty and provide knowledge for the franchisor. Thus, marketing research will help determine what is going right and what went wrong. Research can help the franchisor understand consumers' buying motivations and determine the strengths and weaknesses of the products or services.

When developing a research plan, there are three objectives of market research to remember:

- *Exploratory*: To gather primary information that will better define problems and suggest improvements.
- *Descriptive*: To describe the market potential for a product or the demographics and attitudes of consumers who buy the products.
- *Causal*: To understand the cause and effect relationships in market practices.

The three major functional roles in marketing research are descriptive, diagnostic, and predictive. The descriptive role focuses on gathering and presenting statements of fact. For example, what has been the sales trend for the franchise outlet during the last five years? What are the consumers' attitudes towards the product or service? The second role of research goes beyond description and is diagnostic.

A researcher attempts to explain the data. What was the impact on sales when we introduced the new product? Have sales increased as a result of the new product line introduced? The final function of research would be predictive. Will this individual operate a franchise in a manner consistent with franchise philosophy? Can we predict how a new advertising campaign will affect sales? If we replace the manager, will we improve on employee morale and performance?

The third stage of a marketing research project is implementing the research plan by collecting and analyzing the data. The franchisor must determine the best instrument to use to collect data for the project. Interviews, focus groups, and surveys can all be used to collect data, but the costs, ease, and depth of information vary from one method to another. For example, when conducting research about a franchisee's goals, more in-depth information will probably be required than can be provided by a simple questionnaire.

The fourth step in marketing research is interpreting and reporting the findings where conclusions are drawn to the franchisors. The purpose of this research is not to overwhelm management but provide information which will be useful for making decisions.

THE FRANCHISE DILEMMA

The franchise dilemma is that franchisors have two distinct and different target markets, including:

- The prospective franchisee
- The end consumer.

Unlike most major retail or service outlets, the franchisor does not sell directly to an end consumer but through a franchising organization. The franchises are generally owned and operated by franchisees who actually run, manage, or supervise the outlet stores. It is, therefore, important for the franchisor to first develop a success profile for prospective franchisees.

USING RESEARCH TO CREATE A FRANCHISEE PROFILE

The successful franchisee profile is best developed by devising a preference research model. Such an approach involves research using both personal interviews and a general survey instrument. The personal interview is generally completed with existing successful franchisees. The purpose of the interview is to measure franchisees' attitudes towards their managing experience, as well as general demographic information. The franchisees selected for the study are generally among the top quartile or top third in sales or other success criteria as determined by the franchisors.

In addition to the personal interview, a general survey is made of all franchisees concerning what they believe would be a success profile for franchisees. This general survey is often developed by first using focus groups. The focus group generally includes 6-10 people gathered together for a few hours with a trained interviewer to talk about franchisees, managers, and successful management techniques.

These individuals may be the franchisor's franchisees, but they may also include other managers and owners of different types of businesses, as long as they have some understanding of management and ownership responsibilities. The trained interviewer generally begins with broad questions before moving to specific issues. The interviewer encourages open discussion, hoping that the interactions between the participants bring out actual experiences and recommendations for successful franchisees. These discussions are generally taped for future playback and study.

Once a focus group has determined critical factors that lead to successful franchisees, a survey of franchisees can begin. The purpose of the survey is to obtain primary data concerning attitudes towards the franchisor and the franchise system, and goals of the franchisee, as well as demographics. It is with this information that the franchisor can begin to understand where problems with relationships with their franchisees are occurring.

MARKET RESEARCH

One of the most typical functions of a marketing research programme is to understand the habits and attitudes of customers. This information, when provided to franchisees by the franchisor, can build stronger channel relationships. The customer profile is often conducted through exit interviews as well as a general survey.

The exit interview generally is conducted as customers leave a store. At this time, their attitudes about the product or service are measured. In addition, perception indicators can indicate how individuals felt about the quality of the product or service being sold compared to competitors' offerings. Also, demographic information will provide valuable information about the customer segment currently visiting the franchisee's outlet. With the customer profile for each outlet, franchisors may be able to tailor their franchise programme to better meet the unique needs of each franchisee, hence reducing conflict within the relationship.

Additionally, marketing research may be conducted with focus groups which will allow for greater insights into the end consumer. Focus groups may indicate the strengths or weaknesses of a particular company, product, or service. The focus groups may easily be divided up into heavy users, moderate users, light users, or non-users of a particular franchise outlet. Marketing research can play an integral part in proper decision making in a franchising organization. More and more franchisors are recognizing the benefits of implementing a marketing research programme.

Along with the commonly recognized benefits of marketing research, franchisors should also recognize that marketing research can be an effective tool in building strong relationships with its franchisees.

First, marketing research can help franchisors select franchisees whose goals are consistent with system-wide growth. Second, franchisors can conduct customer research and share the findings with franchisees.

Marketing research provides information which franchisors can use to properly determine the profile for both the successful franchisee and the target customer. When both the franchisee and customer profiles have been developed, the franchisor will be better able to select franchisees who can manage the franchising outlet and serve the customer. Marketing research becomes one tool to reducing unnecessary conflict within franchise channels.

The increasing efforts by marketers to target diverse groups of consumers call for a closer examination of the ethical implications of market segmentation and differentiated marketing. Previous research suggests that marketers and consumers often differ in their perceptions of marketing ethics. Based on contingency theory, this research proposes an integrated framework—which includes the nature of the product, consumer characteristics, and market selection—to analyse the ethical complexities of the marketing exchange. Interactions among these factors lead to various contingencies with different ethical implications for marketing managers and public policy makers. Marketers should assess consumer interests and the ethics of marketing programmes before their implementation

In the last several decades, targeting distinctive consumer segments with differentiated marketing has been a popular strategy among many marketers. The distinctive nature of various consumer groups such as children, the elderly, women, and ethnic minorities has made them attractive market segments. However, market segmentation and targeted marketing have, from time to time, met with tremendous difficulties. Targeting potentially harmful products at vulnerable and disadvantaged consumers such as children, the elderly, and inner-city residents has received negative publicity and been subjected to damaging litigation. The increasing willingness of some large corporations to exploit vulnerable consumers indicates unfair treatment of these consumers and a lack of justice in the marketplace.

On the other hand, cases of discrimination and redlining still occur in certain product and market areas, such as the discrimination directed at minority consumers by insurance and mortgage companies. Disenchanted consumers are increasingly rallying consumer interest groups to put pressure on firms that they consider to be predatory or discriminatory in their marketing practices. Companies such as R. J. Reynolds, Prudential Insurance, and Texaco Oil Company have suffered from tarnished images, consumer boycotts, and court penalties of hundreds of millions of dollars.

Although the modern marketing concept emphasizes its mission to satisfy consumer needs and wants, that promise, in reality, is sometimes lost or misplaced, resulting in outcomes that are not in the best interests of either the customers or society. A lack of understanding of the ethical issues associated with market segmentation and selection has contributed to these problems, which have tremendous social and economic costs. While targeting harmful products at vulnerable consumers has received harsh criticism, restricting the marketing of certain products and labeling some consumers as vulnerable are considered equally troublesome, suggesting that the ethical implications of marketing practices are complicated.

Although researchers have examined the ethics of market segmentation and selection, effort is lacking in synthesizing various issues to provide a holistic understanding of the ethical implications of the marketing exchange. Based on contingency theory, we integrate previous research and propose a three-dimensional framework--which includes the nature of the product, consumer characteristics, and market selection--to analyse the ethics of market segmentation and related marketing strategies. Interactions among these variables lead to various scenarios with different ethical implications for marketing management and public policy making.

Particular attention is given to areas in which marketers and consumers differ in their perceptions of ethical propriety with respect to the nature of the product and market selection. To avoid ethical conflicts in marketing, we suggest that companies consider consumers' interest and assess the ethical implications of their marketing plans before implementation. In the marketing field, market segmentation and the differentiation of the marketing mix have become so prevalent that they are almost synonymous with competitive strategies.

While targeted marketing has largely been beneficial to both consumers and marketers, cases of consumer discontent from time to time raise questions about the ethical implications of market segmentation and targeted marketing, most notably the targeting of harmful products at vulnerable consumers, such as targeting alcohol and cigarettes at inner-city consumers and churning insurance policies to the elderly. Meanwhile, the opposite of targeting--the exclusion of certain consumers from a company's offerings--is just as controversial.

These problems are not unique to marketers of consumer goods. Business service providers and public and nonprofit organizations are not immune to such challenges. Organizations, including educational institutions and government agencies, that cater to certain segments of society have also become the targets of public scrutiny and face the prominent possibility of consumer discontent and legal actions.

While consumers and marketers continue to struggle with these complicated issues and seek viable solutions, researchers have studied the ethics of market segmentation and market selection. Recent studies have focused on the ethical implications of targeted marketing of harmful products at vulnerable consumers.

TARGETED MARKETING

Targeted marketing, as a popular marketing strategy, refers to the concentrated marketing of a product to a segment of consumers due to the attractiveness of the group, in terms of such factors as its size and growth rate. Theoretically speaking, there is nothing inherently wrong with targeted marketing. When marketers promote a product beneficial to a group of consumers, targeted marketing is largely ethical and welcome by consumers.

For instance, some restaurants and hotels have special promotion programmes aimed at children, and indirectly at their parents, with inexpensive toys and special vacation packages. In recent years, however, many cases of targeting potentially harmful products at vulnerable consumers have raised ethical concerns in terms of justice and fairness, such as targeting sweepstakes at the elderly and handguns at women. Even indirect and subtle targeting of potentially harmful products at vulnerable consumers has received criticism, such as targeting children with R-rated movies and using animal characters to promote cigarettes and alcohol.

PRODUCT HARM

Many products in the marketplace, such as wholesome food and medicine, are beneficial or at least beneficial to the target group. However, products that are intended to be beneficial have from time to time been found harmful to some consumers, resulting in"market failures".

For instance, unless a pharmaceutical company tests a new product on everyone (instead of limited clinical trials), it is difficult to determine whether the product will cause any harm. The key issue here is the marketer's knowledge of the product's potential harm, even to a small group of people.

If a company discovers that a small percentage of people may suffer negative reactions from its products and it immediately takes corrective measures, the company may be deemed as acting ethically, believing that it should take responsibility for the incident as part of its business liability.

If the company had prior knowledge of the side effect and it took no remedial measures upon such discovery, it would be considered as intentionally harming the consumers.

Having prior knowledge of the harmfulness of a product and withholding that information have been used as evidence against companies, such as in the Dow Corning case of silicone breast implants and the legal battle against tobacco companies.

However, the evaluation of a product's harmfulness is not always clear-cut. Even the marketing of seemingly beneficial and harmless products can sometimes take an unexpected turn, because such products can be harmful to consumers who have known characteristics. Some products, even though not inherently harmful, can be potentially harmful to consumers due to abuse or misuse.

For instance, targeting alcoholic beverages at poor inner-city consumers is particularly problematic, as this segment already suffers from a greater number of alcohol-related health and social problems than the general population. Thus, ethical evaluations of many products depend on their interaction with consumer characteristics and marketing practices.

VULNERABLE CONSUMERS

Recently, consumer vulnerability has drawn much attention in studies of marketing ethics. In numerous legal cases, the court system in the U.S. has defined vulnerable consumers as a group of people who, due to various idiosyncrasies, are sensitive and susceptible to the potential negative effects associated with using a particular product. For instance, a small group of people, due to their body biology, may suffer side effects from using certain medications.

Moreover, children do not have the same level of knowledge, experience, or maturity as adults to process commercial information. Many elderly people, due to their physical and/or mental conditions, also face challenges as consumers. In addition, some consumers may be prone to addiction or compulsion while others may be disadvantaged due to their social and economic conditions.

Such examples of targeting vulnerable consumers include the targeting of high interest loans of credit cards to consumers with poor credit histories and less financial sophistication. Thus, product harmfulness is heightened for consumers at a risk or disadvantage, and poses ethical concerns.

DISCRIMINATION

While much recent research has examined the ethics of targeting harmful products to vulnerable consumers, some researchers continue to investigate the opposite case: the exclusion of certain consumers in marketing. Discrimination in marketing refers to the practice of denying access to products to a group of consumers due to their racial or ethnic background, age, gender, or other characteristics. Discrimination may also have to do with discrepancies in product quality and variety, and the terms of exchange such as pricing and payment method. Since the 1960s, many studies have investigated pricing discrimination against the poor and minority consumers. Despite the tremendous progress in the last few decades, discrimination and redlining have persisted in areas such as housing, mortgage lending, and financial services. Here, the important gauge points are the equivalence or proportionality in marketing intensity and the disparate treatment of consumers.

Existing studies have made significant progress in conceptualizing the ethics of market segmentation and differentiated marketing. In several important respects, their findings have elevated our understanding of the ethical implications of the marketing exchange. First, these studies emphasize that the concern with the ethical consequences of marketing practices in the United States is well grounded in the fundamental beliefs in social and moral equity, fairness, and justice for all people.

Second, researchers have emphasized the ethical risks inherent in market segmentation and differentiated marketing. Dunfee, Smith and Ross suggest that due to the boundary-expanding nature of marketing, ethical concerns may arise in certain situations when marketers cross such boundaries. For instance, marketing products to youth that are intended for mature consumers, such as alcoholic beverages, leads to different ethical evaluations.

According to Star, the social discontent and ethical concerns associated with marketing stem from the functional limitations of the marketing concept. Contemporary marketing practices often rely on market segmentation, which divides consumers into submarkets based on characteristics such as age, gender, race, and nationality. To provide unique values to consumers and to avoid head-on competition, each market segment may require a unique marketing mix. Thus, segmentation and subsequent marketing practices often imply differential albeit unequal treatment of various consumer groups, and raise ethical concerns.

Third, extant studies have found that ethical concerns or dilemmas often arise when a marketer's perception is in conflict with that of consumers or the public. In fact, the customer interface represents the area of marketing in which conflicts in the perception of the ethical propriety of marketing practices frequently occur. Thus, marketers must study the perception of the public when formulating marketing strategies in order to avoid potential misunderstandings.

NEED FOR INTEGRATION

The review of literature reveals several deficiencies in the existing research. First, previous studies have focused on one or another issue of the marketing exchange and lack the integration of disparate perspectives. While much recent research has concentrated on targeting harmful products at vulnerable consumers, the practice of excluding some consumer groups is equally controversial. However, discussion of these two topics has largely been separate in the existing literature.

Theoretically speaking, these two issues are closely related to each other because discrimination and redlining have as much to do with market segmentation and selection as targeted marketing, but in opposite directions. Moreover, insufficient effort has been expended to examine the many positive marketing activities that target disadvantaged groups, such as charity, corporate sponsorship, and public service programmes.

Although the existing literature condemns unethical marketing practices, it has not offered consistent guidelines to determine exactly when the marketing of certain products to some consumers becomes ethically problematic. There are also questions as to whether some consumer characteristics are the causes of vulnerability or simply its correlates. For instance, when controlling for education and income, the effect of vulnerability among ethnic groups may be less pronounced.

Thus, several authors have cautioned against the use of race- or ethnicity-based vulnerability. Government intervention to restrict marketing may have some merit, and at the same time finds strong opponents among those who support the free market and free speech. Often, consumer advocates and industry spokespeople find themselves in deadlocks in the heated debate about the ethics of market segmentation and selection.

Recent studies have revealed complex interactions among products, consumer characteristics, and marketing practices in ethical evaluations of marketers, and have recommended that marketers should tread carefully when venturing into certain market segments. Despite these research efforts and the many theories of marketing ethics that have emerged, a framework for comprehensive analysis of the ethical implications of market selection and related marketing strategies remains elusive.

Noting the manifold issues and perspectives, several researchers stress the need for a normative theory of ethics about market selection and marketing strategies. A more integrated approach to analyzing the ethical implications of the marketing exchange can help chart a coherent discourse on these critical issues and a better understanding of marketing ethics for managers and public policy makers.

A CONTINGENCY APPROACH

Since the early 1980s, businesses and researchers have devoted much attention to the ethical implications of business practices, which has signaled the arrival of the "ethics era" in which consumer sovereignty is given more emphasis than marketers' interests, and there has been a decline of the caveat emptor assumption. A creative variety of frameworks and theories has been developed within the marketing ethics literature, such as the general theory of marketing ethics, moral decision-making theory, and social contracts theory. These theories typically rely on one or more of the classical theories, such as Kantian ethics and perspectives of rights, duties, and justice.

Existing theories of marketing ethics have developed various "tests" to examine the ethicality of marketing practices. Some of these ethical tenets are related to product harm, consumer characteristics, and market selection. For instance, Laczniak and Murphy's framework of ethical reasoning elaborates several such tests. The motive test asks whether the intent of the contemplated action is harmful. The consequence test establishes whether any major damage to people or organizations will result from the contemplated action.

The justice test asks whether the proposed action leaves another person or group less well off, and whether this person or group is already a member of a relatively underprivileged class. Although these tests have pragmatic appeals, there may be conflicting responses to their questions due to multiple stakeholders in the marketing exchange.

Despite the frequent ethical problems that involve consumers, few theories have considered the role of consumer perceptions in marketing ethics. The exceptions include the consumer sovereignty test and distributive justice. The consumer sovereignty test (comprising three dimensions) requires marketers to examine whether consumers are vulnerable or disadvantaged, perhaps due to age, education, of income.

Marketers need to establish whether a target market has the capability to understand the benefits and risks associated with a product, has sufficient information to judge whether their expectations for purchases will be fulfilled, and has the choice of going elsewhere. The theory of distributive justice suggests that the unequal and differential treatment of consumers, when it leads to detrimental effects on their well-being, particularly of those who are disadvantaged, may result in a perceived lack of fairness and justice.

However, most normative theories of marketing ethics that have emerged in the last two decades have focused on either defining the fundamental principles of ethics to develop guidelines, or on the moral development and ethical reasoning of managers.

Although this approach undoubtedly has merit in assisting marketers to reflect on the ethics of their decisions, it is less satisfactory in providing definitive ethical evaluations of specific practices. Meanwhile, others have proposed universal codes of marketing ethics, such as those of the American Marketing Association, that deal with various ethical issues in different marketing areas such as advertising, sales management, and marketing research. However, they do not specifically address the complex interplay among consumer characteristics, product types, and marketing strategies.

CONTINGENCY THEORY

Dunfee, Smith, and Ross contend that the pluralistic approaches in current marketing ethics research reflect the inadequacy of general normative theories of marketing ethics to handle the rich and complex context of the marketing function. Thompson argues that making decisions on the basis of the context-independent general principles while disregarding the contextual details can result in socially irresponsible, and even detrimental, consequences.

Although most marketing ethics theories, particularly the classic utilitarian (teleological) theories of ethics, consider the interests of various parties and complex situations, they have been criticized as too abstract and general to provide adequate guidance for managers. Due to the great number of ethical issues inherent in customer relationships and marketing functions, one set of universal rules or principles of ethics may be too simplistic to apply to all marketing situations.

To provide clear guidelines for ethical decision-making in marketing, several researchers have emphasized the role of various cultural, organizational and environmental factors in ethical evaluations of marketing activities, including the contextualist approach and the social contracts theory. The contextualist approach stresses that the sources of ethical concerns for marketers come from the interests of different stakeholders and affect the evaluation of the ethical justifiability of a particular marketing activity.

According to the social contracts theory, ethical concerns arise when one community's legitimate norms (e.g., those of marketers) conflict with those of another community (e.g., consumers or the larger society). Therefore, a theory that considers different stakeholder interests and distinguishes various marketing scenarios and their ethical evaluations would be the most fruitful.

One of the theoretical approaches well suited to analyzing complex interactions among various dimensions of a phenomenon is contingency theory. Rooted in organizational behaviour research by Herbert Simon, contingency theory deals with the effect of interactions among various organizational and environmental factors on performance outcomes. According to Zeithaml, Varadarajan, and Zeithaml, contingency theory emphasizes the importance of situational influences on the management of organizations and questions the existence of a single best way to manage. Contingency theory contributes to practical management by encouraging managers to:

- Identify important contingency variables that distinguish between contexts,
- Group similar contexts based on the variables, and
- Determine the most effective solution for each group.

The contingency approach is particularly helpful in analyzing the ethics of marketing activities. Based on contingency theory, we reject the possibility that a single set of criteria can determine marketing ethics in all circumstances.

Instead, we examine the interactions between the contextual variables and emphasize the analysis of the ethical implications of specific scenarios. As the most frequent ethical conflict that faces marketing managers involves attempting to balance corporate interests against those of consumers, a useful framework should first examine how a marketer's perception of marketing ethics interacts with that of the consumer, who is as an integral and indispensable party to and a key stakeholder of the marketing exchange process.

Marketing is often considered as an exchange between marketers and consumers that aims to satisfy consumer needs and maximize the return on investment for shareholders. Thus, there is an inevitable and omnipresent tension between marketers' interests and those of consumers. This conflict forms the basis for different positions on the ethics continuum of marketing practices.

However, placing consumers' interests against those of marketers on the ethics continuum may be too simplistic, because it may imply that marketing is a zero-sum game and reject the possibility of a win-win outcome. Both marketers and consumers may form their perceptions of the ethics of specific marketing scenarios according to ethical principles such as rights, justice, fairness, and equity. In many cases, marketers and consumers agree on the ethical evaluations of certain marketing scenarios and raise no ethical concerns. For instance, both groups agree in principle that providing beneficial products to consumers is ethical as well as desirable and responsible.

However, existing research has documented significant gaps between marketers and consumers in their ethical philosophies as well as ethical evaluations of specific marketing practices. From time to time, a marketer's perception may conflict with that of the public, which results in ethical concerns and, perhaps, disapproving behaviour by the public such as protests and boycotts.

For instance, consumers may believe that reduced pricing for medications such as AZT for the critically ill is ethical and laudable, but manufacturers may consider it unacceptable and unethical based on their obligations to shareholders. In other cases, while consumers perceive targeting harmful products at vulnerable consumers as unethical and exploitative, some marketers consider any restriction of marketing as an infringement of their legal right to free speech under the First Amendment. Thus, consumers' ethical evaluations of a particular marketing situation may be negative, while the marketer's perceptions are based on different moral priorities.

Given such complexities, this research integrates previous studies, applies contingency theory to analyse the ethical implications of the marketing exchange, and explains how ethical concerns arise in various marketing situations based on the agreement of lack thereof in the perceptions of ethics between marketers and consumers. Several recent studies have used two-dimensional frameworks and focused on the targeting of harmful products at vulnerable consumers.

In our three-dimensional model, we include three concepts as the contextual variables, i.e., the nature of the product, consumer characteristics, and market selection. Furthermore, we examine three categories in each variable. In the following sections, we define each of the three contingency variables and discuss how interactions among these dimensions distinguish between various contexts, and how ethical conflicts or dilemmas may arise under certain circumstances.

NATURE OF THE PRODUCT

The nature of a product is an important contingency

variable that affects the ethical impacts of marketing practices. Although there is a continuum of product benefits and harms, products can generally be classified as beneficial, harmful due to abuse/misuse, of inherently harmful. Many products, such as wholesome food and medications, provide a variety of benefits to consumers, including physical, functional, emotional, and social benefits.

On the other hand, products such as cigarettes and other tobacco products are inherently harmful, and it is unethical to market such products. Moreover, some products are benign to sophisticated consumers but may be harmful to others due to abuse of misuse. Alcoholic beverages, for instance, may be beneficial or benign for mature adults who use them responsibly, but are harmful to under-aged youths. The greater the harm of a product, the harsher will be the criticism of marketing practices promoting the product.

Although a physical product itself may be beneficial, the nature of a product's marketing correlates may be unethical. Besides bodily harm, consumers may suffer from economic harm such as the loss of benefits due to deceptive pricing, and psychological harm such as disrespect, humiliation, and a sense of powerlessness that is caused by fraudulent and irresponsible marketing practices such as product defects and deceptive investment schemes.

In addition to consumer discontent and social censure, awareness of the economic impact and psychological effects of unethical marketing practices on consumers has become more important, given the penalties for both economic and psychological damages under the current corporate sentencing guidelines.

CONSUMER CHARACTERISTICS

Existing studies suggest that consumer capability is the most salient factor in ethical evaluations of marketing practices. Clearly, consumer capability/vulnerability is multifaceted. Consumers need mental, physical, and economic abilities to effectively execute a marketing exchange. At times, however, some consumers may lack one or another type of ability to the extent that they can not make informed decisions about a product or benefit from it. Lack of such capabilities may lead to some degree of vulnerability.

In general, a consumer's ability falls on a continuum that runs from being sophisticated to being vulnerable. Here, we group consumers into three categories: sophisticated, at-risk, and vulnerable. Many consumers may be sophisticated due to maturity, education, prior experience, or professional background. Some consumers are considered at-risk when they are prone to addiction or compulsion even though they may possess basic skills and capabilities. These consumers also include those who are socio-economically disadvantaged. For instance, some consumers may be at greater risk if they already suffer from a disproportionate occurrence of harmful behaviour patterns, including those who may not be able to control their own behaviour, such as alcoholics, smokers, or other drug addicts. Moreover, some consumers may be vulnerable due to personal characteristics such as age, education, and physical and mental health.

MARKET SELECTION

With respect to any consumer group, a marketer basically has the following options: to target the group with a unique marketing mix, to market to the group via integrated mass marketing, or to exclude the group from marketing programmes. While integrated mass marketing remains a popular strategy for many products, targeted marketing is considered more effective for companies to gain competitive advantages.

When a product has universal benefits, market selection involves decisions about the inclusion of one group of consumers or the exclusion of another, and may be perceived as unequal or unfair. When a company's marketing programmes exclude certain consumers inadvertently or by design, consumers may have limited choice or incur higher switching costs of going elsewhere.

According to the contingency approach, each of the dimensions discussed above is important in ethical evaluations of marketing activities. However, none of the dimensions alone can determine the ethicality of a marketing situation. The degree of negative or positive ethical evaluation depends on the interaction among the three dimensions. To inform marketing practice and provide ethical guidelines, a normative framework of marketing ethics should enable decision-makers to make better moral judgments that are applicable to specific and often complex situations. Thus, analyzing the specific contingencies of a given marketing activity would be more helpful for assessing the ethical evaluations of marketing than would offering a limited set of general prescriptions.

CONTINGENCIES AND ETHICAL IMPLICATIONS

They have different ethical implications due to differences in perceptions between marketers and the public based on the universal principles of equality, fairness, and justice. Each block provides a unique interaction of the three factors. The white block indicates positive or non-negative ethical evaluation. Black indicates negative ethical evaluation that marketers should avoid.

These two areas often represent situations of universal agreement about ethical practice. The gray block suggests caution due to potential ethical concerns and represents the area in which there are different ethical perceptions and hence disagreement on the ethicality of marketing practices. Based on contingency theory, we describe the following seven common marketing practices and discuss their ethical implications.

Mainstream Marketing

First, companies may market beneficial products to sophisticated and at-risk consumers through targeted marketing or mass marketing, as in the blocks from A1B1C1 to A2B1C2. In these cases, companies will receive positive evaluations from consumers and should proceed. Ringold suggests that marketing transactions with equal participants (i.e., capable and informed consumers) are typically regarded as fair and ethically sound. While many consumers are sophisticated and will not be considered at-risk or vulnerable under normal circumstances, this is not to say that unethical practices would not occur in these situations. Unethical practices may affect capable consumers (who do have bounded rationality and are sometimes vulnerable to unscrupulous marketing) as evidenced by numerous fraudulent investment schemes. As well, in the case of silicone breast implants, many sophisticated consumers fell victim to unethical practices.

Positive Targeting and Social Marketing

Second, targeting and integrated marketing to vulnerable consumers is positive as long as the products are beneficial, and may be meaningful in some product categories (blocks A1B1C3 and A2B1C3). Children, due to their body size and stage of mental development, are often targeted with beneficial products and features such as specially formulated medications, safety caps for medication containers, and educational toys.

Targeting beneficial products at vulnerable consumers mostly receives positive ethical evaluation. Based on the concept of social responsibility, many businesses provide beneficial products to vulnerable and disadvantaged consumers, such as charitable donations to those who are weak and poor. Social marketing programmes also target beneficial products at these consumers, such as free vaccinations for the children of poor families, and low interest rate mortgages for households with limited income. Marketers should receive positive evaluations from the public for these goodwill practices that contribute to consumer well being.

Marketing Potentially Harmful Products

Third, the gray areas are the most challenging for marketing organizations and merit caution (blocks A1B2C1 and A2B2C2 are coded gray). Targeting products that are harmful due to abuse/misuse at any consumers is generally a questionable practice. The marketing of such products to consumers who are already at risk and disadvantaged is even more problematic. For instance, targeted marketing of malt liquor of higher alcohol content to inner-city minorities by some companies led to severe criticism and consumer boycotts due to the disproportionate consumption of such products and more frequent alcohol-related health problems among these groups. Therefore, when marketing products that are potentially harmful due to abuse or misuse, it is important that marketers examine the characteristics of their target markets and use extreme caution to avoid marketing such products to consumers at risk or a disadvantage.

Predatory Marketing

Fourth, marketing of harmful products to any consumer group is unethical, as indicated by blocks from A1B3C1 to A1B3C3, which are coded black. Targeted marketing is appropriately met with intense public scrutiny and even consumer boycotts when products are detrimental to consumer well being and lead to health and social problems. For this reason, the marketing of cigarettes has been curtailed in many countries.

Negative ethical evaluations are even stronger when harmful or potentially harmful products are targeted at vulnerable consumers, because companies are considered to be preying on the weaknesses of such consumers (A1B2C3 and A2B2C3). Because cigarettes and other nicotine products are detrimental to consumer health, it is even more problematic when they are targeted towards the vulnerable (e.g., young children).

Discrimination and Redlining

Fifth, the exclusion of certain consumer groups, particularly vulnerable and disadvantaged consumers, from access to beneficial products is unethical because such actions limit consumer choices. It is also illegal to exclude any consumer groups from marketing programmes specified under U.S. laws such as the Fair Lending Act and the Fair Housing Act. However, segmentation and targeted marketing often mean the inclusion of some groups and exclusion of others, which results in inadvertent redlining. For instance, when marketers launch new products or expand into new geographical markets, they often select business locations based on criteria such as demographic statistics, the cost of doing business, and crime rates. Naturally, companies choose areas where consumers are affluent and the cost of doing business and crime rates are low, thus bypassing poor communities.

Hence, the conflict between consumer interest and corporate objectives poses an ethical dilemma. A number of banks, insurance underwriters, and mortgage companies, such as American Family Insurance and Chevy Chase Bank, faced lawsuits by minority consumers for alleged discrimination and redlining under the Fair Lending Act and the Fair Housing Act.

Although the companies admitted no wrongdoing in the consent decrees, it is the effect, not the intent, of these marketing practices that causes cries of "redlining" from those experiencing the lack of equity and fairness. In such cases, marketers need to ensure that their market selection and marketing activities are based solely on valid business criteria and consumer characteristics that are not directly related to factors such as age, gender, or ethnicity.

Demarketing

However, not all exclusive marketing practices are unethical. Some exclusion is not only positive but also necessary to protect consumers (blocks A3B3C1 to A3B2C3). Many social marketing programmes are designed to dissuade consumers from adopting harmful of potentially harmful products, such as alcoholic beverages, cigarettes, and other addictive drugs.

In addition to efforts by government agencies and civic organizations, many corporations also sponsor and participate in these programmes. To encourage responsible consumption of alcohol products and to prevent drunk driving, Anheuser Busch Company, for instance, sponsored a consumer education programme that included advertising, designated drivers, and free taxi rides. These demarketing efforts play a critical role in educating and protecting consumers and minimizing the potential harm to them.

Reverse Discrimination

Finally, targeting beneficial products at vulnerable and disadvantaged consumers is not entirely without controversies, especially when other groups are not included in these programmes. While marketers may face discontent from disadvantaged consumers—who are the traditional targets of discrimination—lawsuits also come from the non-target segments.

For instance, long distance telephone carriers and airlines have designed special promotions for certain nationality groups who make more calls or travel more often to certain destinations. These and other affinity programmes give the target consumers special rates or discounted fare. Members of other ethnic groups, who are automatically disqualified, may cry discrimination for exactly the same reasons as traditional victims of discrimination.

Although most reverse discrimination cases have to do with employment-related issues, some are directly related to social marketing programmes that are designed to assist the disadvantaged. In a number of cases, white entrepreneurs have charged. Small Business Administration with reverse discrimination, in that these non-members of protected groups have been discriminated against precisely due to their ethnic background.

Thus, marketers need to pay attention to the perceptions and responses of non-target groups, who may derive different meanings from such marketing efforts. Any targeted marketing programme should be fair and sensitive to non-target consumer segments.

ETHICAL IMPLICATIONS ASSESSMENT

A society has many avenues to ensure that marketing will benefit rather than harm consumers. Consumer activism, legislative activity, and government enforcement help to guard against unethical conduct by corporations. While"obey the law" and legal clearance of marketing programmes by companies may be necessary for ethical conduct, they are apparently not sufficient.

Ethical decision-making for businesses requires companies to act in"enlightened self-interest" to ensure the integrity of their marketing programmes. Today, many marketing organizations have adopted the consumer-centreed business philosophy, making"customer satisfaction" their number one priority. If marketers truly subscribe to that mission, then consumers' interests should be given more weight in resolving potential ethical conflicts in marketing decisions.

To fully respect consumer sovereignty and uphold the principles of fairness and justice, we recommend that marketers conduct an"ethical implications assessment" of their marketing programmes. Companies could establish a Marketing Ethics Committee, which includes consumer voices, to conduct such exercises and make the decisions about marketing practices for a given product. Companies can take the following steps using the contingency framework that we have outlined.

First, prior to the launch of products or marketing programmes, companies need to study the perceptions of consumers, including those of non-target segments, in terms of consumer capability, nature of the product, and marketing strategies. In the second stage, the above dimensions should be examined simultaneously to determine the type of contingencies that the company may be facing and whether"ethical clearance" should be given to the proposed marketing programme.

If the marketing programme lands in the black zone, the committee should recommend stopping or aborting the programme. The committee can endorse a marketing programme only if it appears in the white zone. If the results point to the gray zone, the committee should recommend further revisions to avoid potential pitfalls. Finally, only when all negative or questionable contingencies are cleared should the committee give the final seal of approval. By so doing, marketers have a better chance of ensuring that their marketing programmes will benefit consumers, thereby creating benefits as well for the company in a win-win situation.

COMPARISONS WITH OTHER THEORIES

Normative theories of marketing ethics provide a basis for moral deliberation by practitioners and others of the many complex and often troubling ethical issues in marketing. As researchers and practitioners search for meaningful guidelines amidst complex interactions, the contingency approach provides a plausible framework for analyzing the ethics of the marketing exchange and can help sound decision-making in marketing and public policy. The contingency framework considers the parties to and objects of the exchange, and related marketing strategies. Thus, it is comprehensive and enables systematic analyses of the ethical implications of marketing across multiple dimensions and various scenarios. Analyzing the specific contingencies can help generate concrete guidelines for ethical decision-making.

Compared with the general normative theories of marketing ethics, the contingency approach is more flexible in analyzing the ethical implications of various situations based on the three contextual variables. Meanwhile, unlike ethical or moral relativism, the contingency approach stresses the universal principles of equity, justice, and fairness in analyzing specific scenarios, hence remains truthful to a normative theory.

Because the framework considers different products and consumers with different levels of capability as well as the differential impact of marketing on consumers, the perspective of justice provides the most compelling basis for the ethical evaluations of marketing practices. In a sense, the contingency framework serves as a bridge between the principled approach and the utilitarian perspective of marketing ethics, which seem to be incommensurable with each other.

The framework can be expanded in greater detail and applied to other consumer characteristics such as age, income, and education. Products may range from economic models to premium models with enhanced features. The ethical implications of other components of a marketing mix, i.e., distribution, promotion, and price, should be explored. Distribution strategies, for instance, may range from exclusive to selective of intensive distribution.

These marketing practices, when interacting with consumer characteristics and product categories, may lead to ethical concerns. Given the increasing efforts by marketers to target various consumer segments with unique marketing mix strategies, the ethical implications of a marketing campaign based on market segmentation and differentiated marketing warrants systematic and rigorous examination before implementation.

Several practical implications of this approach can be derived. First, to encourage ethical behaviour, companies and industry organizations have established codes of ethics of relied on self-regulation, consumer ombudsmen, or external audits. However, these efforts alone are not enough to eliminate unethical conduct.

Many companies have been relatively passive in examining their positions in marketing ethics and are still operating according to traditional business models and processes that do not consider consumer interests and the ethical implications of their actions as critical issues. The ethical implications of marketing activities often remain afterthoughts, and are yet to be systematically incorporated into management decision-making. The contingency framework can facilitate this preemptive approach to ethical decision-making.

To integrate ethics into a firm's planning and strategy formulation processes, marketers should study consumers' ethical evaluation of their marketing programmes. An"ethical implications assessment" is necessary before the implementation of a marketing programme. In addition to financial, market, and competitive objectives, marketers should include consumers' interest and ethical integrity as important criteria for management decision-making.

Furthermore, ethics must be coordinated throughout the

marketing planning process from product development, market selection, advertising and promotion to implementation and evaluation. Given the competing priorities facing companies in their decisions, finding an ethically sound synergy among product types, consumer characteristics, and marketing strategies is as complicated as solving a crossword puzzle--it has to make sense in all dimensions.

The contingency approach can also serve as a framework for public policy makers to analyse the ethical implications of marketing activities. While public policy makers can take the initiative to protect consumers and promote good corporate citizenship, they also face tremendous challenges and complex dilemmas when making critical decisions, such as restricting certain types of advertising or the availability of some products. For instance, should the legislature make marijuana and similar drugs legal for distribution because they may benefit one group of people?

Regulators need to consider the benefit accruable to one group of consumers as well as the potential harm to another. As Laczniak and Murphy put it,"weighing the concerns of multiple stakeholder groups... becomes the essence of appropriate ethical decision-making... [and] the root of the complexity of such decision-making". While public policy makers should certainly protect the rights of marketers in a free market economy, they must also consider the interests of consumers, particularly those with various degrees of vulnerability, such as children and the elderly.

Future studies of marketers' and consumers' ethical evaluations of the various marketing contexts would help to validate the contingency approach. How consumers form their perceptions of marketers' intentions and make ethical evaluations of marketing will be a fruitful avenue for future research. Researchers may focus on specific situations that would raise ethical concerns for marketers, especially variations in consumers' ethical evaluations of marketing practices (e.g., advertising and distribution) for different market segments (e.g., children and the elderly). In addition to consumer capability, information and choice are two important criteria to determine marketing ethics under the principle of consumer sovereignty. Thus, how companies present information about their products, and how they handle potentially damaging information, may help to shed some light on the ethical decision-making of marketers.

The increasing diversity of the marketplace in the U.S. has significant and complex implications for marketing practices. As more companies are concentrating their resources on the most desirable and profitable segments of consumers, they inevitably exclude other groups.

Consumer and marketing research should use more representative samples of diverse groups to examine consumer responses to marketing programmes, including the responses of non-target markets. In the foreseeable future, the United States will be a marketplace in which all Americans are minorities, and that will pose special challenges and opportunities for both private companies and public organizations. Determining when segmentation and differentiation may lead to perceptions of discrimination or reverse discrimination will continue to be a focal issue for future research.

This research focuses on marketing and consumers in the United States. Cross-country and cross-cultural research would help to understand marketing ethics in a broader context. In the past, multinational corporations experienced difficulties with their marketing strategies in other countries, as in the case of Nestle's baby formula. Despite adverse public attitudes at times, multinationals and local companies used a variety of techniques and media to promote their products to consumers in developing countries without close scrutiny of the consumer perceptions of such practices.

In light of increasing globalization, more companies are crossing national boundaries to produce and market their products in other countries. Thus, how the local communities of the global marketplace evaluate the marketing practices of multinational corporations has become an important subject of investigation.

8

Working Quality and Quantitative Techniques

MORALE

Of the thousands of public sector agencies at the state and local levels of government, a very small percentage are in economic, political or legal situations that allow them to be a community, state or regional leader in terms of employee pay. Organizations that have adopted a pay or compensation philosophy as a market pay "leader" have generally come, not surprisingly, from the private sector.

It is generally difficult for public sector organizations to adopt "market leader" pay philosophies since such a philosophy would generally translate to a greater tax burden for citizens, who can be expected to harbor reservations, at the very least, about the "appropriateness" of a "market leader" pay strategy. In fact as the 1976 ICMA text on personnel administration notes, "It is doubtful that taxpayers would tolerate public employees' salaries exceeding those of their counterparts in industry."

Many or perhaps most public sector organizations are committed to a commonly accepted concept in public compensation theory, that is, to pay public sector employees at "prevailing wages." The concept of "prevailing wages" is generally defined as wages that match the external market for similar positions at approximately the 50th percentile, meaning approximately fifty per cent of similar jobs in the community have lower pay ranges and fifty per cent have higher pay ranges. This pay policy is also commonly referred to as a "neither lead nor lag" policy. Community pay leaders, on the other hand, go well beyond the prevailing wage concept and typically set employee pay ranges in the 70th through 90th percentile.

Are there advantages that accrue to an organization that adopts a community pay leader position? According to Edward E. Lawler III, there may indeed be certain payoffs from pursuing a policy of being a top-notch payer. Lawler notes that, "Such organizations as IBM and Hewlett-Packard have for a long time recognized that if they pay well, they will have very little turnover and be able to pick from among a large number of job applicants." Would public sector organizations realise any benefits if they were able to be community pay leaders? The following paragraphs describe the results for one public sector pay leader.

A PAY LEADER FROM THE PUBLIC SECTOR

As noted earlier, few public agencies are in a position to be a community, state or regional pay leader, but there are exceptions. One such exception is the municipal water utility in Denver, Colourado. This non-union organization, with approximately one thousand employees, is an independent agency of the City and County of Denver with its own policy-setting Board of Commissioners and its own civil service system.

The agency serves water to over 900,000 people, more than a quarter of the state's population, through 250,000 water taps. Its operations are funded through water rates and other fees. Essentially the agency is financially structured as an enterprise fund that does not rely on property taxes. For several years this agency has been a community pay leader and has positioned itself at approximately the 75th percentile of market pay rates. The pay leader position has brought expected and unexpected results for the agency. As expected and as private firms have found, being a pay leader has resulted in a very low employee turnover rate and has contributed to a large pool of applicants vying for job openings. The most important unexpected result has been the low level of job satisfaction expressed by the agency's employees.

In January of 1992 the Mountain States Employers Council, Inc. conducted an Employee Opinion Survey of all agency employees. The response rate for the Survey was 86 per cent. The Mountain States Employers Council had previously conducted many similar Employee Opinion Surveys in the Denver metropolitan area, therefore, the Survey used at the agency contained 34 standard questions with norms developed by previous participants.

Mountain States Employers Council divided the standard questions into several categories comparing the agency scores for these categories to the community norms that had been developed. Two points of interest emerged from an examination of the responses. First, in the category of pay-related questions and the category of benefit-related questions the agency's employees were significantly more positive about pay and benefits than the community norms. Second, in the category of questions related to employee job satisfaction the agency's employees were significantly less satisfied than the community norm.

The agency employee scores that deviated most significantly from the norms in a positive direction involved three questions pertaining to satisfaction with pay increases, satisfaction with the benefit package, and satisfaction with communication about how pay is determined. The most negative scores in terms of community norms concerned the job satisfaction issues of: opportunities for advancement; promotion policies; supervisory support of quality work; and consistency in the application of disciplinary procedures.

Analyzing the survey findings using the perspective of the community norms developed by the Mountain States Employers Council, the following points are clear.

On the positive side:

- Agency employees were generally very satisfied with their base pay and pay increases.
- Agency employees were generally very satisfied with the agency's benefit package.

- Agency employees were generally very satisfied with the equipment they use to perform their jobs.
- Employees felt that they had good job security.
- Employees were satisfied that the agency was concerned about employee safety.
- Employees felt that the amount of work they were asked to do was fair and reasonable.
- Employees felt personally responsible for "top quality" in performing their jobs.

On the negative side:

- Agency employees generally did not feel that promotions were given to the most qualified people.
- Agency employees generally felt very dissatisfied with their chances for advancement. On a related question employees believed that there were not enough opportunities for advancement for qualified people.
- Many agency employees strongly believed that "who you know" was important for getting ahead.
- Many agency employees believed that there were "communication problems" between different sections or departments of the agency.
- Comments from a significant number of employees expressed dissatisfaction with the agency's Affirmative Action policies.
- Employees were also dissatisfied with Human Resources and felt the HR department was not responsive to their needs.

Perhaps the best indicator in the Survey of overall employee job dissatisfaction involved the question "Would you accept a job at another company, doing the same job as you do here, with the same pay and benefits?" Compared to the community norm, a significant percentage of agency employees indicated they would accept another job elsewhere for similar pay and benefits.

The problem is, of course, when you work for a community pay and benefits leader and have excellent job security, the opportunity for finding a similar job elsewhere in the community (the Denver metropolitan area) with equivalent pay and benefits is very, very limited.

THE RAMIFICATIONS OF BEING A PUBLIC SECTOR PAY LEADER

The single most important consequence attributable, at least in large part, to the pay leader position has been a very low turnover of employees. The agency's turnover rate has persistently been in the two to four per cent range during much of the past decade. While on the surface a very low employee turnover rate may appear desirable, it does have its drawbacks.

For example, there are very few advancement opportunities for employees and intense interest focuses on any promotional opportunity that comes along. Additionally there is extreme unhappiness in the general employee population with any programme or civil service policy, rule or procedure seen as adversely affecting an individual's chances for advancement.

In fact a very low turnover rate makes programmes to ensure workforce diversity much more difficult. In an article on turnover in the federal government, Gregory Lewis notes that although turnover may be costly for the federal government, it does have some benefits such as bringing "fresh blood" into an organization and in offering promotion opportunities for lower level employees. Lewis further notes that "turnover was key to improving the representation of women and minorities."

Exacerbating the problem of job dissatisfaction at the agency is a relatively recent change from a "promote from within" policy to one of more aggressively opening promotional opportunities to both internal and external candidates. Making job satisfaction problems worse, the agency's excellent Tuition Reimbursement Programme and nearby college campuses have motivated many employees to obtain college degrees.

A highly educated workforce may appear to be a good thing, but there now appear to be a substantial number of lower level agency employees with college degrees and very little opportunity for advancement. This combination of factors is predictably one that does not lead to a workforce with high levels of job satisfaction. There are, of course, factors other than lack of job advancement opportunities that may be contributing to employee job dissatisfaction. Three of these potential factors are discussed below.

INTERNAL EQUITY

Individuals in organizations generally make two distinct types of pay comparisons when deciding whether or not they are being fairly compensated: internal comparisons of pay with other employees within the organization and external market comparisons. Of these two types of pay comparisons Lawler believes that external market comparisons are most important.

He states that "If external comparisons are poor, an individual will usually leave as soon as he or she can find another job." While external equity is important so is the matter of internal equity. If employees perceive a lack of internal equity between jobs within the organization, morale problems will undoubtedly result.

The agency in question was aware of potential problems resulting from a lack of internal equity among the 300 plus job classifications. A point-factor job evaluation system was used to determine the relative worth of each job and to place each job in the job worth hierarchy in the appropriate pay grade. Wallace and Fay note that job evaluation systems cannot, by themselves, be the answer to all problems concerning internal equity.

Agency employees have, however, expressed very little dissatisfaction with the job evaluation system in use or with comparisons of internal equity in general. There is a job audit procedure in place at the agency so that employees or their supervisors are able to request an audit of any job they perceive incorrectly placed in a particular pay grade.

ORGANIZATION CULTURE

The agency does appear to be undergoing a gradual shift in organizational culture away from a conservative, "family-like" culture. The agency's culture has also reflected at least some of the elements of an "entitlement culture" described by Judith Bardwick. In an entitlement culture, according to Bardwick, employees have so much security that they don't have to earn their rewards.

Bardwick states that "When people don't have to earn what they get, they soon take for granted what they receive. The real irony is that they're not grateful for what they get. Instead, they want more. It is the terrible cycle of entitlement." Pay policies can certainly contribute to a culture of entitlement. For example, if employee pays above the market rate of pay (in this case step 4 is the market step in a 9 step pay system) is not pay at risk, then it may be seen as an entitlement by employees.

It would seem logical that as the agency attempts to modify its organizational culture the level of employee job satisfaction could decrease. However, since job security at the agency remains intact, the impact of a gradual change in organization culture cannot fully explain the level of employee job dissatisfaction evidenced in the Employee Opinion Survey.

MANAGEMENT PROBLEMS

The Survey results could also be indicative of "poor" management at the agency. A large majority of respondents, however, did rate the agency Manager as doing a "satisfactory" or a "better than satisfactory" job. The agency also has a strong supervisory training programme which should, in theory at least, lead to a well-trained management team. Undoubtedly some agency managers are less than fully competent, but on the whole, the Survey does not indicate an incompetent management team or a poorly managed organization.

In reflecting on the Survey results and recent interviews with an unscientific sample of agency employees, it must be concluded that the position of being a community pay leader was primarily responsible for accomplishing the agency objective of low employee turnover. It must also be concluded that the accomplishment of the low turnover objective contributed significantly to the general level of employee job dissatisfaction the agency has experienced.

As the agency's Manager of Employment and Compensation during this period, I can certainly state that a higher-than-average level of employee job dissatisfaction was not the intended consequence of maintaining a position as a community pay leader. Edward Lawler has perhaps summed it up best when he noted that "There is some research that suggests that high pay rates can lead to high motivation.

However, motivation from this source seems to be very short-lived. Most individuals quickly decide that they deserve whatever pay rate they receive and do not try to perform better in order to deserve it." With no evidence of increased productivity at the agency it must be concluded that Lawler's words ring true for this public sector agency.

NEW DIRECTIONS - A PROPOSAL

Few public sector organizations, as noted earlier, appear to be in a position where they can legitimately pursue a strategy of being a community, state or regional pay leader. In any case, the sense is that a pay leader policy does not tend to produce desired results in public sector organizations with elabourate civil service systems. This is a private sector practice that does not translate well to the public sector.

In many organizations, both public and private, the overall aging of the workforce translates to fewer promotional opportunities and more "plateaued" employees. Even for public agencies that have the ability to pay and can afford to be community pay leaders, it would seem that there needs to be a better understanding of nonmonetary career motivators.

Being a pay leader attracts people who are interested in high pay, though they may not necessarily interested in "public service." Perry and Wise note that utilitarian incentives, such as individual pay and benefits, are not likely to be critical determinants of outputs when individuals identify with the mission of the organization, provided that the incentives are maintained at satisfactory or prevailing levels.

They also conclude that "Public service motivation is

likely to be positively related to an individual's organizational commitment. Individuals who are highly committed are likely to be highly motivated to remain with their organizations and to perform."

A failure to attract employees who have at least some level of public service motivation may result in employees who are dissatisfied with their jobs in a public sector agency. The goal is not, it seems, to create utilitarian incentives that attract the "best and the brightest," but rather to provide a level of incentives to attract the best and the brightest who are motivated, for any number of reasons, to serve in the public sector. One key point is that whatever the cause of public sector employee job dissatisfaction, the use of high levels of utilitarian incentives (pay and benefits) alone cannot overcome it. The result will be highly paid, dissatisfied employees. There is a direction that seems to make more sense for public sector employers then simply "upping the ante" to compete with high flying private enterprises.

As Romzek notes, utilitarian incentives are important to attract and retain high quality public employees, but they are not enough. Of equal importance is the necessity for public agencies to foster a commitment by employees to the values and mission of the organization. In addition to fostering employee commitment to organizational values and mission, public sector managers can impact employee turnover rates through a better understanding of career dynamics.

Thomas Barth argues that the Career Anchor Theory developed by Edgar Schein could provide a conceptual framework for understanding career motivation. Barth notes that "Schein maintains that as employees progress through their careers they are engaged in a process of self-discovery that reveals career anchors of which they were not initially aware. If this discovery leads to the conclusion that they are not in the right job or career path, employees have three principal options:

- The employee leaves;
- The employee stays and represses the career anchor;
- The employee and the organization restructure the

job or provide other opportunities within the organization to tap the employee's career anchor."

MEASURES

JOB PERFORMANCE

An overall measure of job performance was obtained by combining self-rated performance scores and supervisory rating scores from two 7-point Likert-type scales assessing performance in terms of quality and quantity. The integration of supervisory and self-ratings of performance may constitute a more objective measure of performance.

When asked to evaluate the quality of their performance, participants were requested to respond to the following two questions: (a) How would you rate the quality of your own performance in your job? and (b) How do you think your supervisor would rate the quality of your performance? Two identical questions were asked about the quantity of performance.

Supervisors were simply requested to assess the performance of their subordinates in terms of quality and quantity, separately. Both participants and supervisors rated the quality and quantity of performance using the following 7-point scale: excellent = 7; very good = 6; good = 5; average = 4; fair = 3; poor = 2; and very poor = 1.

The aggregate score of the two scales (quality and quantity) from both respondents and supervisors was taken as the measure of performance. The self-rated score of quality for each participant was determined and averaged with that provided by the supervisor. The same procedure was used to calculate the score of quantity of performance. The quality and quantity scores were then averaged to arrive at the overall measure of performance.

Data analyses yielded a significant (p [less than].01) relationship between the self-rating of quality of performance and the supervisory rating of quality of performance (r =.32). A weaker but still significant (p [less than].01) relationship was also found between the two measures of quantity obtained in the same fashion (r =.19). After the integration took place, the correlation between the performance dimensions of quality and quantity (r =.62) was significant at p [less than].01.

JOB SATISFACTION

The Job Descriptive Index or JDI was the means for measuring job satisfaction. The JDI assesses five dimensions of job satisfaction:

- Satisfaction with work,
- Satisfaction with pay,
- Satisfaction with promotions,
- Satisfaction with supervision,
- Satisfaction with coworkers.

The JDI has been described as one of the most carefully developed scales for measuring job satisfaction. Its continuous validity makes it an index with the greatest acceptance among management researchers.

ORGANIZATIONAL COMMUNICATION

The Roberts and O'Reilly organizational communication questionnaire was used to measure organizational communication as operationalized in this investigation. Roberts and O'Reilly reported that their instrument had desirable psychometric properties. The mode of data collection was a pre-tested questionnaire. The readability of the instrument was tested via a graduate class in statistics at a local university.

After securing copies of the lists of employees of the participating firms, a numbered copy of the questionnaire was distributed to each member of the sample. Each individual was told that the number on the questionnaire would be used only as a means for matching statistical data and that no one else would see the number except the researchers. This number was later used to identify each respondent's supervisor.

The internal mail system of each of the participating organizations was used to deliver the research instrument and introductory letter. The completed questionnaire was to be mailed to the Department of Management of a nearby university. A total of 316 employees completed and returned their questionnaires. Of this total, seven had removed their identification numbers. Two others were improperly answered. Consequently, 307 usable questionnaires were left. The task was then to ask supervisors to rate respondents' performance. With the assistance of the Director of Human Resources of each participating firm, the immediate supervisor of each respondent was identified using the number assigned to each returned questionnaire. Seventy-nine supervisors were identified. Again, the internal mail systems of the participating organizations were used to deliver the performance scale and related materials to the supervisors. The job performance assessments were to be returned to the University.

One supervisor of three individuals did not respond. Also, during the two weeks taken by supervisors to assess participants' performance, one of the original respondents resigned while another was dismissed. These two respondents were therefore dropped from the list of actual respondents. The elimination of these last five participants left a total of 302 participants for whom supervisors provided performance evaluations. This total of respondents represents a 49.35% rate of return.

With a span of control ranging from 2 to 8 subordinates, none of the 78 supervisors who participated in the assessment of respondents' performance accounted for a disproportionate amount of the data collected. Further, no event, project, or activity was detected that might have influenced participants' performance during the period that lapsed between the time respondents completed the research instrument and the time supervisors conducted the performance assessment.

STATISTICAL ANALYSES

Hypothesis one was tested by noting whether the correlation between job performance and job satisfaction was positive and significant. Hypothesis two, concerning the expected moderating influence of organizational communication on the performance-satisfaction relationship, was assessed by moderated regression analysis.

Since the theoretical argument that performance leads to satisfaction has been adopted in this investigation, satisfaction was treated as the criterion, with performance and organizational communication as pre-dictors.

Following Saunders and Zedeck, the basic regression equations included the interaction or cross-product of the predictors. This addition provided the following three regression equations for evaluation purposes, where Y is satisfaction, X is performance, Z is organizational communication, and a to d are constants.

- $Y = a + bX$
- $Y = a + bX + cZ$
- $Y = a + bX + cZ + dXZ$

Zedeck states that "if Equations 2 and 3 are significantly different from Equation 1, but not from each other, then the variable [the suggested moderator, organizational communication] is an independent predictor and not a moderator variable". Therefore, for this investigation, a moderator exists only in situations in which a predictor-moderator interaction term accounts for a significant amount of criterion variance after both predictor and moderator variables have entered the regression equation.

Moderated regression analyses were conducted for all organizational communication dimensions investigated. This effort also provided any predicting effects detected. The mean, standard deviation and median for each variable show high variability in participants' responses. To compare them, all scores were reduced to a seven-point Likert-type scale. More will be said about this variability in the discussion section.

Results for the relationship between performance and satisfaction show a direct and moderate relationship between them. Thus hypothesis is accepted. The highest correlation was between performance and satisfaction with work, and the lowest between performance and satisfaction with coworkers.

Assessment of the potential moderating effect of organizational communication on the relationship between job performance and job satisfaction, the central issue of this study, begins with a review of the correlation matrix. In general, the communication dimensions (potential moderators) show varying degrees of co-variance with both the independent variable (job performance) and the dependent variable (job satisfaction).

Of the seven communication dimensions investigated, directionality of communication (upward, downward, and lateral) and communication load (underload and overload) would appear to have the greatest potential as moderators. Conversely, trust in superiors, perceived influence of superiors, accuracy of information, desire for interaction, and satisfaction with communication appear to have the least potential.

The aim of this investigation is to explore the moderating influence of organizational communication on the relation-ship between job performance and job satisfaction. The study also re-examines the association between these last two variables. Results show that job performance has a direct, weak-to-moderate relationship with job satisfaction.

It also replicates the finding by Petty, McGee, and Cavender, who used meta-analysis (a statistical technique which accumulates findings across studies). Thus, previous and current findings suggest, once again, that changes in one of the two variables (performance or satisfaction) may only weakly or moderately influence the other.

Organizational communication received weak support as a moderator of the relationship between performance and satisfaction. Only 2 of the 50 interactions were significant: satisfaction with work was influenced by the interaction of accuracy of information and satisfaction with pay was impacted by the interaction of lateral communication. The correlation between performance and satisfaction with work was greater for individuals scoring high in accuracy of communication than for participants scoring low. It appears that appropriate and accurate information may enhance both performance and satisfaction with work.

This finding implies that individuals receiving proper, correct, and clear information may perform adequately, which in turn may give rise to positive feelings about their jobs, or vice versa. Supervisors might be able to promote adequate levels of job performance and job satisfaction among their employees by providing them with appropriate and accurate information.

On the other hand, individuals scoring low in lateral communication reflected a greater correlation between performance and satisfaction with pay than respondents scoring high. This finding suggests that employees subjected to low levels of lateral communication may be less inclined to make nondesirable comparisons of job related features, including pay, which in turn may negatively affect both performance and satisfaction with pay.

Such a suggestion may be especially true in places where salary or wage differences are not perceived as adequate or consistent. Ample opportunity for lateral communication may increase the possibility of perceiving pay as inadequate when compared with the contribution being made. Further investigation of this speculation may prove beneficial to management practitioners. When considering the very weak support that organizational communication received as a moderator, one question arises. Why was such support so weak? Two factors may help answer this question. First, this investigation used a very rigorous statistical procedure for detecting moderating effects - moderated regression analysis. Second, many of the communication dimensions studied received strong support as independent predictors.

Thirty of the 50 relationships investigated showed that the integrated communication dimensions acted as predictors. Zedeck suggests that moderators are very difficult to find when the change after adding the proposed moderators as a predictor to the regression equation is high to start with. In this study, the change resulting from the addition of a proposed communication moderator to a particular model as a predictor was equal to or greater than.13.

Because of this phenomenon, six of the seven communication dimensions investigated - trust in superiors, influence of superiors, accuracy of information, desire for interaction, satisfaction with communication, and communication load - received strong support as predictors of job satisfaction.

In addition, job satisfaction was treated as the dependent variable when conducting the moderated regression analysis. Considering that virtually every dimension of communication investigated related to both job performance and job satisfaction in the same manner, it is then plausible to suggest that the main effects of organizational communi-cation are compatible with two of the central theories in the literature concerning the causal relationship between performance and satisfaction:

- Job performance - [greater than] job satisfaction,
- Job satisfaction - [greater than] job performance.

For example, high levels of accuracy of information may lead to high levels of performance, and successful performance may promote job satisfaction. The sequence may also function in reverse. High levels of accuracy of communication may lead to high levels of job satisfaction, and favourable perception of satisfaction may enhance job performance.

What appears evident from this research is that regardless of the direction of the job performance-job satisfaction relation, organizational communication appears to be an important predictor of both variables. Thus, communication can be an effective tool that practitioners may use to enhance these two dimensions. Downs, Clampitt, and Pfeiffer argue this same point.

Finally, some methodological observations should be made. Admittedly, measuring organizational and individual constructs by aggregating self-reported data from research participants, as was done in this study, is controversial. The basic criticism "is essentially that researchers using survey methods have simply combined disparate perceptions into fallacious averages".

In this study, however, the researchers are confident that the investigated measures of job performance, job satisfaction, and organizational communication are meaningful. First, these findings about the relationship between job performance and job satisfaction mirrored those of previous studies, suggesting that the variables analysed were not just fallacious or spurious averages.

Second, a graphical analysis of residuals yielded properties that suggested that the residuals were independent, had zero mean, had a common variance, and followed a normal distribution. Therefore, these reasons provide strong support for the point that the data collected adequately represent the constructs investigated in this study.

That organization communication received strong support as a predictor of job satisfaction and weak support as a moderator of the job performance-job satisfaction relationship is the major conclusion of this research. Although some relationships did not prove significant as hypothesized, communication did prove to be important in organizational functioning nonetheless.

And such importance undergirds the need for the continued investigation of communication in organizations. These findings, combined with the use of rigorous statistical procedures and the research design of relating communication to significant output variables, make this study unique in its contribution.

But more research into these complex relationships is needed by business communication scientists. Downs, Clampitt, and Pfeiffer note that, "Perhaps the opportunities are rare, and the choices few, when both objectives [performance and satisfaction] can be achieved through a single communication strategy". Business communication researchers might consider alternate strategies and combinations of strategies in future research investigations.

First, this study did not take a skills approach in defining communication and relating it to performance-satisfaction outputs. Perhaps an investigation of writing (letters and reports) and speaking in the context of this research effort might prove interesting. If specific skill levels are examined, they should be approached within the proper research context, design, and process for meaningful results to occur.

Moreover, this study used only organizational communication as defined by Roberts and O'Reilly. Other constructs of communication such as the Communication Satisfaction Questionnaire or the Interpersonal Communication Relationship Inventory could be used to determine the moderating effects of communication. Or, to determine the influence of the supervisor's interpersonal capacity on the performance-satisfaction relationship, the Index of Interpersonal Communication Competence could be used.

The various definitions of communication in the literature call for several research possibilities to analyse the performance-communication-satisfaction trilogy in organizations. The opportunities for additional research are exciting, to be sure. Business communication researchers need to tap these opportunities as they investigate the complex and dynamic process of communication in organizations.

Index

N

O

P

Q

R

S